DESIGNING TALL BUILDINGS
STRUCTURE AS ARCHITECTURE

THE FIRST OF ITS KIND, *Designing Tall Buildings* is an accessible reference that guides the reader through the fundamental principles of designing high-rises. Each chapter focuses on a theme central to tall-building design, giving a comprehensive overview of the related architecture and structural engineering concepts. Author Mark Sarkisian, PE, SE, LEED® AP, provides clear definitions of technical terms and introduces important equations, gradually developing the reader's knowledge. Later chapters explore more complex applications, such as biomimicry. Projects drawn from SOM's vast portfolio of built high-rises, many of which Sarkisian engineered, demonstrate these concepts.

This book considers the influence of a particular site's geology, wind conditions, and seismicity. Using this contextual knowledge and analysis helps determine what types of structural solutions are best suited for a tower on a specific site, and allows the reader to conceptualize efficient structural systems that are not only safe, but also constructible and economical. Sarkisian also addresses the influence of nature in design, urging the integration of structure and architecture for buildings of superior performance, sustainability, and aesthetic excellence.

DESIGNING TALL BUILDINGS

STRUCTURE AS ARCHITECTURE

MARK SARKISIAN

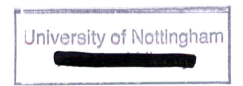
 Routledge
Taylor & Francis Group

NEW YORK AND LONDON

To the Architects and Engineers of Skidmore, Owings & Merrill LLP

First published 2012
by Routledge
711 Third Avenue, New York, NY 10017

Simultaneously published in the UK
by Routledge
2 Park Square, Milton Park, Abingdon, Oxon OX14 4RN

Routledge is an imprint of the Taylor & Francis Group, an informa business

Cover image courtesy of SOM © Tim Griffith
Project: Poly International Plaza, Guangzhou, China

Library of Congress Cataloging in Publication Data
Sarkisian, Mark P.
Designing tall buildings : structure as architecture / Mark P. Sarkisian.
    p. cm.
    Includes bibliographical references and index.
1. Tall buildings--Design and construction. I. Title.
NA6230.S27 2011
720'.483—dc22

                                                  2011004868

Consulting editor: Laurie Manfra
Acquisition editor: Wendy Fuller
Project manager: Laura Williamson
Production editor: Alfred Symons
Designer and typesetter: Alex Lazarou

Printed and bound on acid-free paper by TJ International Ltd, Padstow, Cornwall

ISBN13: 978-0-415-89479-1 (hbk)
ISBN13: 978-0-415-89480-7 (pbk)
ISBN13: 978-0-203-80659-3 (ebk)

1006233810.

CONTENTS

FOREWORD

ONE OF THE central challenges of the 21st century is designing intelligent forms of human settlement. In the last 200 years, the global population grew from 1 billion to 6.9 billion. In less than four decades, the global population will reach 9 billion. Our highly consumptive pattern of development that relies upon an inexhaustible supply of arable land, water, and energy cannot be sustained. Buildings and transportation today create two-thirds of the carbon in our atmosphere. Where we place our buildings, the way we build them, and the way in which we move between them are the major causes of climate change. The future of our planet is contingent upon our ability at the beginning of this new millennium to create cities of delight—urban environments that are dense, compact, and highly livable.

A key to achieving this is an American invention: the tall building. The experiment in vertical building begun by Jennings, Burnham, Sullivan, and others in Chicago after the Great Fire of 1871 is central to the long-term sustainability of our planet. The earliest examples of this building form, begun in earnest in the 20th century, must now be reconsidered with the understanding that the tall building is not simply an expression of corporate power or civic pride—it must become the very basis of human settlement.

To physically realize compact, dense, and humanistic vertical cities in a future of limited material resources, radical innovations in architecture and structural engineering are necessary. Mark Sarkisian and his colleagues at SOM have embraced this challenge. Their process begins not with columns and beams, but rather with an intuitive understanding of the interrelationship between forces at play. They go beyond the rote, normative process of structural optimization to consider new ways of achieving holistic building efficiencies in which the essence of architectural form becomes part of the solution.

Mark advocates an intuitive and organic understanding of structure and form in SOM's design studios. To support this nuanced approach, he and his colleagues have assembled extraordinary new tools to achieve innovation. These systems combine computational analysis with visualization models similar to those used by scientists to quickly visualize organic form and behavior at the molecular level. This work is undertaken in a collaborative, multidisciplinary environment, much like the "group intelligence" model of today's collaborative sciences.

Mark and his colleagues have extended this collaborative model beyond SOM's professional studios to the academic studio. SOM's engagement with leading universities in many disciplines over the years has yielded constant reciprocal benefits—inspiration and research—for the firm and the academy. This book, *Designing Tall Buildings*, is a direct result of the engineering and architecture class at Stanford University that Mark began with architect Brian Lee in October 2007. It is a book that will no doubt provide inspiration as well as practical guidance to students, professors, practicing architecture and engineering professionals, and design devotees.

Craig W. Hartman, FAIA
January, 2011

INTRODUCTION

THIS BOOK IS meant to illuminate the design process for tall building structures with fundamental concepts and initial considerations of the site developed into complex solutions through advanced principles related to natural growth and the environment. The author's goal is to give a holistic description of all major considerations in the structural engineering design process. Specific examples of work developed at Skidmore, Owings & Merrill LLP (SOM) are used for each step in the process.

The work within this book represents decades of development by the architects and engineers of SOM. As an integrated practice, its work is the product of a close design collaboration—leading to many innovations and particularly in tall building design. Pioneering structural systems, including adaptation to form, material efficiencies, and high performance, have resulted from this work.

The catalyst for this book came from the need for a structural engineering curriculum in an SOM-led Integrated Design Studio class at Stanford University. The goal was to teach architectural and structural engineering design, in parallel and with equal emphasis, focused on tall building design challenges which included complex programmatic and site considerations. Each chapter in this book was developed as a class lecture, focused on a particular subject in the design process.

The book begins with a select history of tall buildings, the inspiration behind their designs, as well as some early analysis techniques for their design. The site is considered for geotechnical, wind, and seismic conditions. Forces from gravity and lateral loads, including wind and earthquakes, are described with load combinations required for design. Multiple codes are referenced to offer different approaches to the calculation of loads on the structure.

The language of the tall building structure is described to provide a better understanding of major structural components and overall systems. Framing diagrams from buildings are included to demonstrate the use of major structural building materials such as steel, concrete, and composite (a combination of steel and concrete). Attributes of tall structures are described, including strength and serviceability—building drift, accelerations, and damping—among other things. Tall building characteristics such as dynamic properties, aerodynamics associated with form, placement of materials, and aspect ratios are also described.

Suggested structural systems based on height and materials are reviewed. These systems result in the greatest efficiency through a minimum of material when considering gravity and lateral loads. Inspirations from nature through growth patterns and natural forms are considered for the development of more advanced ideas for tall building structural systems.

Natural behavior is contemplated through the consideration of structures behaving mechanically rather than statically when subjected to load—particularly in seismic events. Correlations to mathematical theories such as the Fibonacci Sequence and genetic algorithms, as well as the use of emergence theory in structural design, are considered. Finally, and perhaps most importantly, effects on the environment, including embodied energy and equivalent carbon emissions, are contemplated.

DESIGNING TALL BUILDINGS

STRUCTURE AS ARCHITECTURE

CHAPTER 1
PERSPECTIVE

THE FIRE OF 1871 devastated the city of Chicago but created an opportunity to re-think design and construction in an urban environment, to consider the limits of available, engineered building materials, to expand on the understanding of others, and to conceive and develop vertical transportation systems that would move people and materials within taller structures.

In the late 1800s technological advancements led to the development of cast iron during the United States' industrial revolution. Although brittle, this material had high strength and could be prefabricated, enabling rapid on-site construction. The first occupied multi-story building to use this technology was the Home Insurance Building located in Chicago. Built in 1885, with two floors added in 1890, it was 12 stories tall with a height of 55 m (180 ft). Though it has since been demolished, it is considered the first skyscraper.

Chicago in Flames—The Rush for Lives Over Randolph Street Bridge (1871), Chicago, IL

The Chicago Building of the Home Insurance Co., Chicago, IL

FACING PAGE
Willis Tower (formerly Sears Tower) (1974), Chicago, IL

The Monadnock Building, Chicago, IL

The 16-story Monadnock Building located in Chicago and constructed in 1891 used 1.8 m (6 ft) thick unreinforced masonry walls to reach a height of 60 m (197 ft). The structure exists today as the tallest load-bearing unreinforced masonry building. The 15-story, 61.6 m (202 ft) tall Reliance Building built in 1895 used structural steel and introduced the first curtain wall system. Buildings now could be conceived as clad structural skeletons with building skins erected after the frame was constructed. The Reliance Building has changed use (office building converted to hotel), but still exists on State Street in Chicago. Steam and hydraulic elevators were tested for use in 1850. By 1873, Elisha Graves Otis had developed and installed steam elevators into 2000 buildings across America. In 1889, the era of the skyscraper was embraced with the first installation of a direct-connected, geared electric elevator.

Identity and egos fueled a tall building boom in the late 1920s and early 1930s with other urban centers outside of Chicago getting involved. In 1930, the Chrysler Building in New York became the world's tallest, with the Empire State Building soon to follow. Completed in April 1931 (built in one year and 45 days), at 382 m (1252 ft), it surpassed the Chrysler Building by 62.2 m (204 ft). The total rental area in the tower is 195,000 square meters (2.1 million square feet). The most significant feat was the extraordinary speed with which the building was planned and constructed through a close collaboration between architect, engineer, owner, and contractor.

Reliance Building, Chicago, IL

The Empire State Building's contract for architectural services with Shreve, Lamb, and Harmon was signed in September 1929, the first structural steel column was placed on 7 April 1930, and the steel frame was topped off on the 86th floor six months later (the frame rose by more than a story a day). The fully enclosed building, including the mooring mast that raised its height to the equivalent of 102 stories, was finished by March 1931 (11 months after the first steel column was placed). The opening day ceremony took place on 1 May 1931. The structural engineer, H.G. Balcom (from a background in steel fabrication and railroad construction), worked closely with general contractors Starrett Brothers and Eken to devise a systemized construction process.

Three thousand five hundred workers were on site during peak activity. Some 52,145 metric tonnes (57,480 tons) of steel, 47,400 cubic meters (62,000 cubic yards) of concrete, 10 million bricks, 6400 windows, and 67 elevators were installed. The Empire State Building remained the

Beaux Arts Architect Ball (1931),
New York, NY

Skyscrapers of New York,
Linen Postcard (1943)

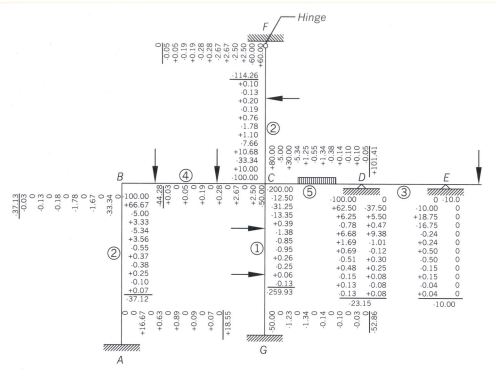

Moment Distribution for Indeterminate Structures

tallest building in the world for 41 years, until the World Trade Center in New York was built in 1972. The Chrysler, Empire State, and World Trade Center buildings were all constructed of structural steel.

The development of more sophisticated hand calculation techniques for structures, including methods developed by great engineers like Hardy Cross, made it possible to analyze, design, and draw structures that could be easily constructed. Urged by the University of Illinois' Dean of Engineering Milo Ketchum, Cross published a ten-page paper entitled "Analysis of Continuous Frames by Distributing Fixed-End Moments" in 1930, showing how to solve force distribution in indeterminate structures, which was one of the most difficult problems in structural analysis.

The Second World War temporarily halted homeland construction because of the need for steel products in the war effort. It wasn't until the late 1950s and early 1960s that interest in tall buildings was renewed. Great architects such as Mies van der Rohe used structural steel to create a minimalistic architectural approach. His notable tall building projects included 860 & 880 North Lake Shore Drive (1951) and 900 & 910 North Lake Shore Drive (1956) in Chicago. Skidmore, Owings, & Merrill LLP (SOM) developed building designs that used structural steel to create long, column-free spans

CLOCKWISE FROM TOP LEFT
Lever House, New York, NY

Inland Steel Building, Chicago, IL

One Bush Street (Formerly Crown Zellerbach), San Francisco, CA

Alcoa Building, San Francisco, CA

Willlis Tower (formerly Sears Tower), Chicago, IL

John Hancock Center, Chicago, IL

allowing for flexible open office spaces while creating a corporate identity through the finished building. These projects included The Lever House—New York (1952), the Inland Steel Building—Chicago (1958), the Crown Zellerbach Building/One Bush Street—San Francisco (1959), and the Alcoa Building—San Francisco (1964).

It wasn't until the late 1960s and early 1970s that considerable new development in tall building analysis, design, and construction were made. The Cray Computer provided the analytical horsepower to evaluate buildings such as the John Hancock Center (1969) and the Sears Tower (1973) located in Chicago. Prefabricated, multi-story, modular building frame construction was used to reduce construction time. Wind engineering, largely developed by Alan Davenport and Nicholas Isyumov at the University of Western Ontario, provided vital information about the performance of buildings in significant wind climates. Geotechnical engineering, led by engineers such as Clyde Baker, provided feasible foundation solutions in moderate to poor soil conditions. Probably the most important contribution was SOM's late partner Dr. Fazlur Khan's development of economical structural systems for tall buildings. His concepts were founded in fundamental engineering principles with well-defined and understandable load paths. His designs were closely integrated with the architecture and, in many cases, became the architecture.

Chestnut-Dewitt Tower, Chicago, IL

Brunswick Building, Chicago, IL

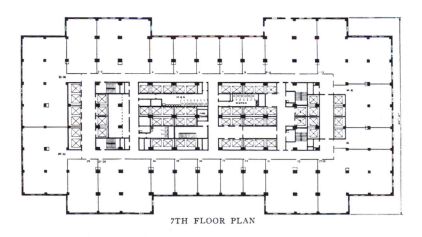

7TH FLOOR PLAN

Empire State Building, Typical Floor Plan, New York, NY

Drawing from *Fortune* Magazine,
September 1930, Skyscraper Comparison

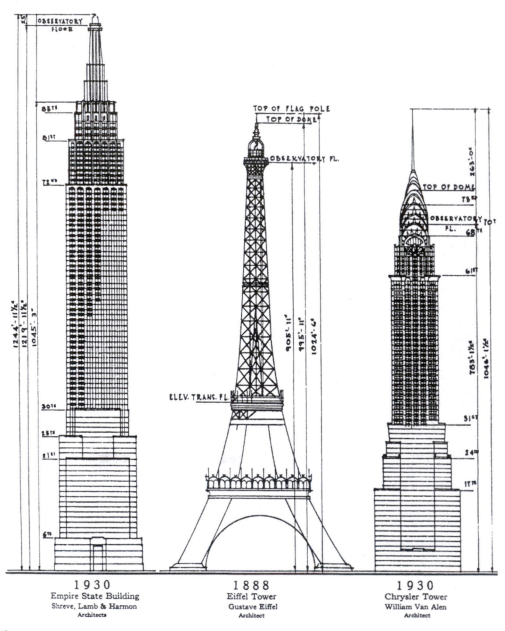

1930
Empire State Building
Shreve, Lamb & Harmon
Architects

1888
Eiffel Tower
Gustave Eiffel
Architect

1930
Chrysler Tower
William Van Alen
Architect

At the same time, SOM also developed tall building structural system solutions in reinforced concrete. An increased understanding of concrete's chemical and physical characteristics combined with consistently higher compressive strengths led to an economical alternative to structural steel in tall building structures. The Brunswick Building (1964) and Chestnut Dewitt Tower (1965) located in Chicago were major structures that used this technology.

Tower height depends on material strengths, site conditions, structural systems, analytical/design capabilities, the understanding of building behavior, use, financial limitations, vision, and ego. Concepts in tall building evolve rather than radically change. Increased understanding of materials combined with greater analytical capabilities have led to advancements. Computer punch cards and Cray Computers have been replaced by laptops with comparable computing capabilities. Many advancements are based on architectural/engineering collaboration.

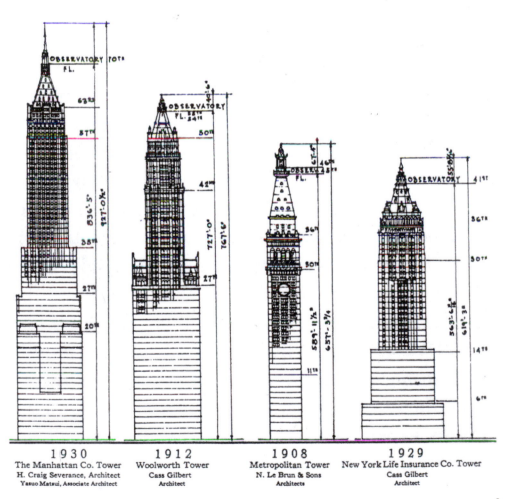

1930	1912	1908	1929
The Manhattan Co. Tower	**Woolworth Tower**	**Metropolitan Tower**	**New York Life Insurance Co. Tower**
H. Craig Severance, Architect	Cass Gilbert	N. Le Brun & Sons	Cass Gilbert
Yasuo Matsui, Associate Architect	Architect	Architects	Architect

CHAPTER 2
SITE

THE PRIMARY SITE CONSIDERATIONS for tower design include the effects of wind, seismic, and geotechnical conditions. The conditions may be code-defined or derived from specific site conditions. The site conditions can be modeled analytically to replicate expected behavior during anticipated events.

Structures 200 m (656 ft) or more in height, even those consisting of reinforced concrete (which has greater mass than structural steel) and located in moderate to high seismic areas, are usually controlled by wind effects rather than seismicity. This by no means relaxes the required ductility, detailing, and redundancy for the structure, but it does mean that the structure is flexible with a significantly long fundamental period of vibration of approximately 5 seconds or more, attracting smaller inertial forces than a shorter structure with a shorter period.

Poor soil conditions, near-fault effects, and potential earthquake intensity must be considered and may change the governing behavior. In fact, certain critical elements within the superstructure may be considered to perform elastically in even a rare earthquake event (475 year event, 10% probability of exceedance in 50 years). For instance, steel members located within an outrigger truss system that are intermittently located within the tower may require considerations for this level of force to achieve satisfactory performance even in an extreme seismic event.

2.1   WIND

2.1.1   General Effects

Direct positive pressure is exerted on the surface facing (windward faces) or perpendicular to the wind. This phenomenon is directly impacted by the moving air mass and generally produces the greatest force on the structure unless the tower is highly streamlined in form. Negative pressure or suction

Wind Flow,
John Hancock Center, Chicago, IL

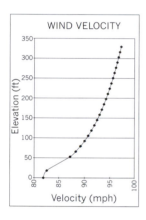

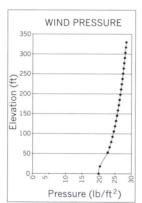

Comparison of Wind Velocity and Wind Pressure

typically occurs on the leeward (opposite face from the wind) side of the tower. Since the winds flow like a liquid, there are drag effects on the surfaces parallel to the direction of the wind. These surfaces may also have positive or negative pressures on them, but it is the drag effect that adds to the general force on the tower. The combination of these three effects generally results in the net force on a tower. However, for very tall or slender structures their dynamic characteristics can produce amplified forces. Across wind or lift motion is common for these structures. In fact, many of these taller structures are controlled by this behavior. This dynamic effect could exist at even low velocities if the velocity of the wind causes force pulses through vortex shedding that match the natural period of vibration of the structure.

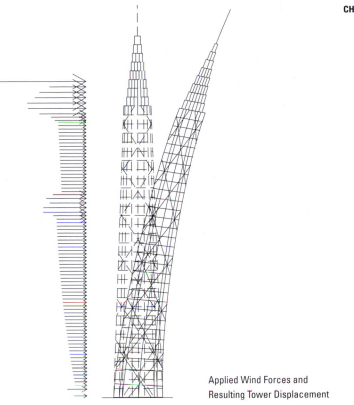

Applied Wind Forces and
Resulting Tower Displacement

Generally, the relationship between wind velocity and pressure is:

$$P = 0.003V^2$$

where,

P = equivalent static pressure on a stationary object (lbs/ft^2)
V = wind velocity (miles/hr)

Although wind conditions are generalized for a given geographic area, the local terrain at a site has a substantial effect on the pressures expected. For instance, the applied pressures expected for a structure located in open terrain are significantly higher than those expected in an urban setting where tall structures may surround the site.

Code-defined wind criteria must be used as the basis for all tall building design; however, these criteria are generally too conservative for the tall tower. Buildings 40 stories or taller should be considered for wind tunnel studies that evaluate realistic structural behavior. These studies result in a rational evaluation of the in-situ wind climate and usually lower base building design forces and provide accurate, local wind effects on cladding and on pedestrians at the ground plane.

Wind tunnel studies should include:

a. proximity modeling/wind climate (detailed modeling of structures within 0.8 km (0.5 miles) of site) and wind environment analysis based on historic data
b. pressure tap modeling of exterior walls
c. pedestrian wind analysis
d. force-balance structural modeling
e. aero-elastic structural modeling (consider for heights over 300 m (984 ft))

2.1.2 Code Requirements

In accordance with the American Society of Civil Engineers 7-10 (ASCE 7-10), Minimum Design Loads for Buildings and Other Structures, Section 6.6 permits rational wind tunnel studies to determine loads on any building or structure in lieu of code formulas; however, many governing jurisdictions require that minimum code-defined loads must be used for strength design. In most instances, in addition to exterior wall design, the rational wind can be used to evaluate the structure for serviceability, including drift and accelerations. This usually leads to a considerable reduction in base building stiffness. According to the 2006 International Building Code (IBC), which includes most of the requirements of ASCE 7-10, base building structural wind load is determined with the following design procedure:

1. Determine the basic wind speed V and the wind directionality factor K_d.
2. Determine the importance factor I.
3. Calculate the exposure category or exposure categories and velocity pressure exposure coefficient K_z or K_h.
4. Calculate the topographic factor K_{zt}.
5. Determine the gust effect factor G or G_f.
6. Determine the enclosure classification.
7. Calculate the internal pressure coefficient GC_{pi}.
8. Calculate the external pressure coefficients C_p or GC_{pf} or force coefficients G_f.
9. Calculate the velocity pressure q_z or q_h as applicable.
10. Calculate the design wind pressure p, where p for rigid buildings is:

$$p = qGC_p - q_i(GC_{pi})$$

and

$$q_z = 0.00256 K_z K_{zt} K_d V^2 I \text{ (lbs/ft}^2)$$

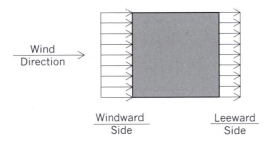

Wind Direction →

Windward Side

Leeward Side

The IBC is based on wind velocities measured 10 m (33 ft) above the ground and on three-second gusts. However, if wind speeds are provided in the form of fastest mile wind speed, the following relationship for wind speeds is given.

Three-second (3s) gust and fastest mile wind speed (mph):

$$V_{fm} = \frac{V_{3s} - 10.5}{1.05}$$

where,

V_{3s} = 3-second gust basic wind speed (available from wind speed maps including those in the IBC 2006)

According to the 1997 Uniform Building Code (UBC), base building design wind pressure is defined as:

$$P = C_e C_q q_s I_w$$

where,

P = design wind pressure
C_e = combined height, exposure and gust factor coefficient
C_q = pressure coefficient for the structure
q_s = wind stagnation pressure at the standard height of 10 meters (33 feet) based on the fastest mile wind speed (highest average wind speed based on the time required for a mile-long sample of air to pass a fixed point)
I_w = importance factor

Most building codes are based on a 50-year return wind event for strength and serviceability design of building structures. There are cases, however, where a 100-year return wind must be considered for design particularly related to structural strength. This increase in design pressure is usually at least 10%.

Jin Mao Tower Wind Tunnel Testing

2.1.3 Rational Wind Considerations

The rational wind can be considered in two components: static and dynamic. Magnitude, direction, and proximity to neighboring structures (both current and future) are important. Frequently, future planned buildings (if known to be part of a master plan at the time of design) could dynamically excite the structure, causing magnification of force levels and requiring a more conservative, yet appropriate design. For instance, in the case of the Jin Mao Building in Shanghai, two taller towers were planned within neighboring city blocks resulting in design forces that were controlled by dynamic effects from vortex shedding of wind from the neighboring structures. This behavior magnified the forces on Jin Mao by 33%.

San Andreas Fault,
Carrizo Plain, CA

2.2   SEISMICITY

2.2.1   Intensity

The intensity of an earthquake is based on a qualitative assessment of damage and other observed effects on people, buildings, and other features. Intensity varies based on location within an affected region. An earthquake in a densely populated area may result in many deaths and considerable damage, whereas the same earthquake may result in no damage or deaths in remote areas. The scale used most commonly to evaluate the subjective intensity is the Modified Mercalli Intensity (MMI) Scale developed in 1931 by American seismologists Harry Wood and Frank Neumann. The scale consists of 12 increasing levels of intensity expressed in Roman numerals. The scale ranges from imperceptible motion (Intensity I) to catastrophic destruction (Intensity XII). A qualitative description of the complete scale is as follows:

Intensity I   Not felt except by very few under especially favorable conditions.

Intensity II   Felt only by a few persons at rest, especially by those on upper floors of buildings. Delicately suspended objects may swing.

Intensity III   Felt quite noticeably by persons indoors, especially in upper floors of buildings. Many people do not recognize it as an earthquake. Standing vehicles may rock slightly. Vibrations similar to the passing of a truck. Duration estimated.

Results of Strong Ground Motion, Olive View Hospital (1971),
Sylmar, CA

Intensity IV   During the day, felt indoors by many, outdoors by a few. At night, some awakened. Dishes, windows, doors disturbed; walls make cracking sound. Sensation like heavy truck striking a building. Standing vehicles rock noticeably.

Intensity V   Felt by nearly everyone; many awakened. Some dishes, windows broken. Unstable objects overturned. Pendulum clocks may stop.

Intensity VI   Felt by all, many frightened. Some furniture moved. A few instances of fallen plaster. Damage slight.

Intensity VII   Damage negligible in buildings of good design and construction; slight to moderate in well-built ordinary structures; considerable damage in poorly built structures. Some chimneys broken.

Intensity VIII Damage slight in specially designed structures; considerable damage in ordinary substantial buildings, with partial collapse. Damage great in poorly built structures. Fallen chimneys, factory stacks, columns, monuments, walls. Heavy furniture overturned.

Intensity IX   Damage considerable in specially designed structures; well-designed frame structures thrown out of plumb. Damage great in substantial buildings, with partial collapse. Buildings shifted off foundations.

Intensity X   Some well-built wooden structures destroyed; most masonry and frame structure with foundations destroyed. Rails bent greatly.

Intensity XI   Few, if any, masonry structures remain standing. Bridges destroyed. Rails bent greatly.

Intensity XII   Damage total. Lines of sight and level are destroyed. Objects thrown into the air.

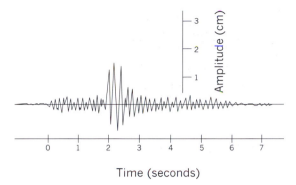

Seismograph Trace

2.2.2   Magnitude

The most commonly used measure of an earthquake's strength is determined from a scale developed by Charles F. Richter of the California Institute of Technology in 1935. The magnitude, M, of an earthquake is determined from the base ten logarithm of the maximum amplitude of oscillation measured by a seismograph.

$$M = \log_{10}(A/A_o)$$

where,

A = measured maximum amplitude
A_o = measured amplitude of an earthquake of standard size (calibration earthquake)
A_o generally equal to 3.94×10^{-5} in (0.001 mm)

The above equation assumes that the seismograph and the epicenter are 100 km (62 miles) apart. For other distances a nomograph must be used to calculate M.

Since the equation used to calculate M is based on a logarithmic scale, each whole number increase in magnitude represents a ten-fold increase in measured amplitude. The Richter magnitude M is typically expressed in whole and decimal numbers. For example, 5.3 generally corresponds to a moderate earthquake, 7.3 generally corresponds to a strong earthquake, and 7.5 and above corresponds to a great earthquake. Earthquakes of magnitude 2.0 or less are known as microearthquakes and occur frequently in the San Francisco Bay Area. The 1989 Loma Prieta Earthquake measured 7.1 on the Richter Scale, with the 1906 San Francisco Earthquake corresponding to 8.3. The largest recorded earthquake was the great Chilean Earthquake of 1960, where a magnitude of 9.5 was recorded.

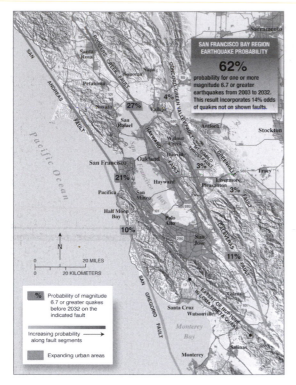

Probability of a 6.7 Magnitude Earthquake or Greater in San Francisco before Year 2032

The seismometer is the detecting and recording portion of a larger apparatus known as a seismograph. Seismometers are pendulum-type devices that are mounted on the ground and measure the displacement of the ground with respect to a stationary reference point. Since the device can record only one orthogonal direction, three seismometers are required to record all components of ground motion (two translational, one vertical). The major movement during an earthquake occurs during the strong phase. The longer the earthquake shakes, the more energy is absorbed by the buildings, resulting in increased damage based on duration. The 1940 El Centro Earthquake (magnitude 7.1) had 10 seconds of strong ground motion and the 1989 Loma Prieta Earthquake (magnitude 7.1) lasted only 10–15 seconds. By contrast the 1985 Chilean Earthquake (magnitude 7.8) lasted 80 seconds and the 1985 Mexico City Earthquake (magnitude 8.1) lasted 60 seconds. There is debate in California, with no consensus, on whether long duration earthquakes can occur. The assumed strong ground shaking for the current 1997 UBC Zone 4 earthquake is 10 to 20 seconds.

Soil Liquefaction (1995),
Kobe, Japan

2.2.3   Energy

The energy released during a seismic event can be correlated to the earthquake's magnitude. In 1956, Beno Gutenberg and Richter determined an approximate correlation where the radiated energy (ergs) is less than the total energy released, with the difference related to indeterminate heat and other non-elastic effects.

$$\log_{10} E = 11.8 + 1.5M$$

Since the relationship between magnitude and energy is logarithmic with the associated factors, an earthquake of magnitude 6 radiates approximately 32 times the energy of one of magnitude 5. In other words, it would take 32 smaller earthquakes to release the same energy as that of one earthquake one magnitude larger.

2.2.4   Peak Ground Acceleration

Peak ground or maximum acceleration (PGA) is measured by an accelerometer and is an important characteristic of an earthquake oscillatory response. This value is frequently expressed in terms of a fraction or percentage of gravitational acceleration. For instance, the peak ground acceleration measured during the 1971 San Fernando Earthquake was 1.25g or 125%g or 12.3 m/s^2 (40.3 ft/s^2). The peak ground acceleration measured during the Loma Prieta Earthquake was 0.65g or 65%g or 6.38 m/s^2 (20.9 ft/s^2).

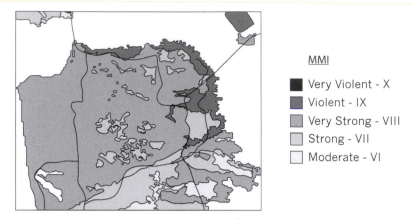

Projected Damage (MMI) in San Francisco from 1989 Loma Prieta Earthquake (M=7.1)

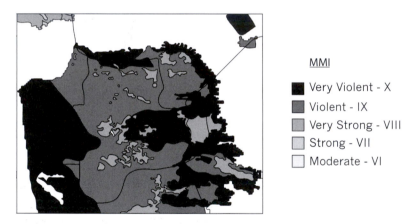

Projected Damage (MMI) in San Francisco from 1906 Earthquake (M=8.3)

2.2.5   Correlation of Intensity, Magnitude, and Peak Ground Acceleration

No exact correlation between intensity, magnitude, and peak ground acceleration exists since damage is dependent on many factors, including earthquake duration and the way that a structure was designed and constructed. For instance, buildings designed in remote locations in developing countries will likely perform much worse than structures designed in major urban areas of developed countries. However, within a geographic region with common design and construction practices, a fairly good correlation may be drawn between intensity, magnitude, and peak ground acceleration.

MMI	PGA	Approx. Magnitude
IV	0.03g and below	
V	0.03g–0.08g	5.0
VI	0.08g–0.15g	5.5
VII	0.15g–0.25g	6.0
VIII	0.25g–0.45g	6.5–7.5
IX	0.45g–0.60g	8.0
X	0.60g–0.80g	8.5
XI	0.80–0.90g	
XII	0.90g and above	

In addition, seismic zones as defined by the 1997 UBC can be correlated to an earthquake's magnitude and peak ground acceleration.

Seismic Zone	PGA	Max. Magnitude
0	0.04g	4.3
1	0.075g	4.7
2A	0.15g	5.5
2B	0.20g	5.9
3	0.30g	6.6
4	0.40g	7.2

In certain limited areas or micro-zones, peak ground accelerations may vary significantly. This variance is primarily attributed to local site soil conditions. During the Loma Prieta Earthquake, peak accelerations measured in San Francisco were generally not greater than 0.09g, but peak accelerations recorded at the Bay Bridge, Golden Gate Bridge and the San Francisco Airport were 0.22–0.33g, 0.24g, and 0.33g respectively. After the Mexico City Earthquake of 1985, micro-zones were incorporated into the rebuilding plan.

2.2.6   Earthquake, Site, and Building Period

Earthquakes release energy in different frequency ranges. The period (or the natural frequency) of a vibration, the time it takes for one full oscillatory cycle, is the characteristic of motion that affects sites and structures. If the site (soil) has a natural frequency of vibration that corresponds to the predominant earthquake frequency, site movement can be greatly amplified through a phenomenon called resonance. Structures located on these sites can experience amplified forces. Soil characteristics such as density, bearing strength, moisture content, compressibility, and tendency to liquefy all may affect the site period.

Theoretically, a structure with zero damping, when displaced laterally by an earthquake, will oscillate back and forth indefinitely with a regular period. As damping is introduced, the motion will eventually stop. The building period is not the site period; however, if these periods are close to one another, resonance could occur with a large magnification of forces that the structure must resist.

2.2.7   Probability of Exceedance and Return Period

Earthquakes are commonly described by the percent probability of being exceeded in a defined number of years. For instance, a code-defined design basis earthquake is typically referred to as having a 10% probability of being exceeded in 50 years. Another way of describing this earthquake design level is through "return period." For this code-defined earthquake (10% probability of exceedance in 50 years) the earthquake is also known as having a 475-year return period or sometimes referred to as a 475 year event. The following describes the conversion between return period and probability of exceedance:

$$RP = T / r^*$$

where,

$$r^* = r (1 + 0.5r)$$
$$RP = \text{return period}$$
$$T = \text{target year of exceedance}$$
$$r = \% \text{ probability of exceedance}$$

Therefore, for a 10% probability of exceedance in 50 years:

$$RP = 50 / 0.10 (1 + 0.5 (0.1)) = 476.2 \approx 475$$

The table opposite includes commonly used probabilities of exceedance and return periods.

2.2.8   Spectral Acceleration

Measured amplitude of an earthquake over time during a seismic event is not regular. It is difficult to determine how a structure behaves at all times during an earthquake consisting of random pulses. In many cases it is not necessary to evaluate the entire time history response of the structure because the structure is likely more affected by the peak ground acceleration than by smaller accelerations that occur during the earthquake. The spectral acceleration is the cumulative result of the interaction of the structure's dynamic characteristics with the specific energy content of an earthquake.

Probability of Exceedance/Return Period Table

Event	r	T	r*	RP	RP (rounded)
63% in 50 years	0.63	50	0.1315	60.4	60
10% in 50 years	0.10	50	0.105	476.2	475
5% in 50 years	0.05	50	0.05125	975.6	975
2% in 50 years	0.02	50	0.0202	2475.2	2475
10% in 100 years	0.10	100	0.105	952.4	975

The spectral acceleration is the maximum acceleration experienced by a single degree of freedom vibratory system of a given period in a given earthquake. The maximum velocity and displacement are known as the spectral velocity and spectral displacement respectively.

The maximum building acceleration is typically higher than the peak ground acceleration, so these values should not be confused. The ratio of the building to peak ground acceleration depends on the building period. For an infinitely stiff structure (period = 0 sec.) the ratio is 1.0. For short period structures in California, considering a 5% damped building and a hazard level (probability of occurrence) of 10% in a 50-year period located on rock or other firm soil, the ratio is approximately 2.0 to 2.5 times the peak ground acceleration (spectral amplification).

Response spectra commonly used in design are developed based on spectral accelerations. These spectra may be site specific or code-defined.

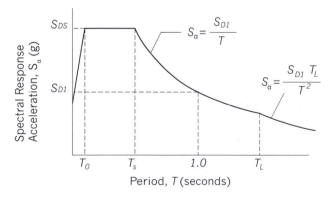

2006 International Building Code Seismic Response Spectrum

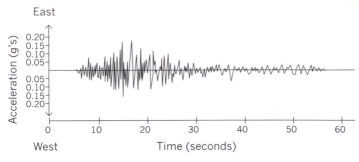

Seismic Time History Record

2.2.9   Design and Maximum Considered Earthquakes

The design basis level earthquake as recognized by the 1997 UBC is based on an earthquake that has a 10% probability of being exceeded in 50 years (approximately a 475 year event). This design level is based on the reasonable likelihood that an earthquake of this magnitude will occur during the life of the structure. At this level of seismicity, the structure is expected to be damaged, but not collapse, and life safety protected.

The maximum considered earthquake represents the maximum earthquake expected at a site. Generally this is an event that has a 2% probability of being exceeded in 50 years (approximately a 2475 year event). Typically structures are designed for stability (collapse prevention) in this earthquake event, but higher performance goals (i.e. life safety) may be required for important/essential facilities such as hospitals or police stations for this extreme event.

2.2.10   Levels of Seismic Performance

- Operational Level (O)
  Backup utility services maintain functions; very little damage.

- Immediate Occupancy Level (IO)
  The building receives a "green tag" (safe to occupy) inspection rating; any repairs are minor.

- Life Safety Level (LS)
  Structure remains stable and has significant reserve capacity; hazardous non-structural damage is controlled.

- Collapse Prevention Level (CP)
  The building remains standing, but only barely; any other damage or loss is acceptable.

High Performance
(Less Loss)

Operational (O)

Immediate Occupancy (IO)

Damage Control
(Enhanced Design)

Life Safe (LS)

Collapse Prevention (CP)

Lower Performance
(More Loss)

Considerations for Seismic Performance

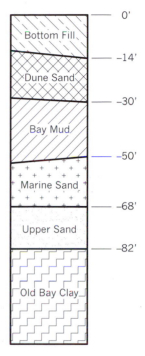

Typical Soil Conditions from South of
Market Street Sites in San Francisco

2.3 SOILS

Geotechnical conditions vary widely for sites of ultra-tall structures. Mechanics of the site soil conditions include stability, water effects, and anticipated deformations. Soil conditions may vary from bedrock to sand to clay, bedrock having the best geotechnical characteristics, with dense sand having similar traits. Sand provides good foundation support since settlement is elastic (associated with initial loading from the structure), but could be difficult to accommodate during construction and could liquefy (complete loss of shear strength) when saturated and subjected to lateral seismic loads. Clay could provide excellent foundation support especially if pre-consolidated, but must be considered for both initial loading effects and long-term creep effects due to consolidation. Clay could prove to be excellent for site excavations.

Spread footings usually prove to be the most cost effective foundation solution, followed by mat foundations. When bearing capacities are low or applied loads are high, deep foundations consisting of piles or caissons are usually required. The following is a general summary considering foundation type.

27

Mat Foundation Prior to Concrete Pour,
Burj Khalifa, Dubai, UAE

2.3.1   Spread or Continuous Wall Footings

Spread or continuous footings are used under individual columns or walls in
conditions where bearing capacity is adequate for applied load. This system
may be used on a single stratum, firm layer over soft layer, or reasonably soft
layer over a firm layer. Immediate, differential, and consolidation settlements
must be checked.

2.3.2   Mat Foundations

Mat foundations are used in similar applications as spread or continuous wall
footings where spread or wall footings cover over 50% of building area. Use
is appropriate for heavy column loads, with the mat system usually reducing
differential and total settlements. Immediate, differential, and consolidation
settlements must be checked.

Pile Foundation,
Jin Mao Tower, Shanghai, China

2.3.3   Pile Foundations

Pile foundations are used in groups of two or more to support heavy column
or wall loads. Reinforced concrete pile caps are used to transfer loads from
columns or walls to the piles. Pile foundations provide an excellent solution
for poor surface and near-surface soil conditions. This foundation system
is a good solution for structures in areas susceptible to potential soil liquefaction. Piles are generally 20–50 m (65–164 ft) long below the lowest basement. Pile capacity is typically developed by skin friction, but end bearing
may also be considered. Piles are usually designed to resist lateral loads
(due to wind or seismic) in addition to vertical load. Bending on piles may
be considered with heads fixed or pinned into pile caps. Piles typically consist of steel or concrete for tower structures (although timber could also
be used). Corrosive soil conditions may require concrete (precast) to be
used. H-piles in structural steel and 355 mm x 355 mm (14 in x 14 in) or
406 mm x 406 mm (16 in x 16 in) precast piles are common. Open steel
pipe piles have been used in conditions of dense sand and extremely high
applied loads.

Caisson Construction,
NBC Tower at Cityfront Center, Chicago, IL

2.3.4 Caisson Foundations

Cast-in-place reinforced concrete caissons typically have a diameter of 750 mm (30 in) or more and may either be straight-shafted or belled. Bell diameters are typically three times the shaft diameter. Caisson foundations provide an excellent solution for poor surface and near-surface soil conditions. The capacity of this system is usually based on end bearing. End bearing of caissons is commonly founded in stiff clay (hardpan). Installation of caissons is very sensitive to soil conditions. Voids in shafts or bells are quite possible due to local soil instability during installation. Concrete may be placed under bentonite slurry to prevent soil instability during installation. The length of the caisson shaft usually varies from 8 to 50 m (26 to 164 ft).

Slurry Wall Construction,
Jin Mao Tower, Shanghai, China

Slurry Wall Construction,
Harvard University Northwest Science Building,
Cambridge, MA

2.3.5 Basement/Foundation/Retaining Walls

Basement/foundation/retaining walls can be used in any soil condition, but usually require controlled, engineered backfill behind the walls. Where permanent water conditions exist, waterproofing is required. Slurry walls, cast under a bentonite slurry, provide temporary soil retention and permanent foundation walls. Bentonite caking at the exterior provides permanent waterproofing. Slurry walls are installed in panels, usually 4.5 m (15 ft) long, with shear keyways existing between panels. Reinforcing typically does not cross panel joints.

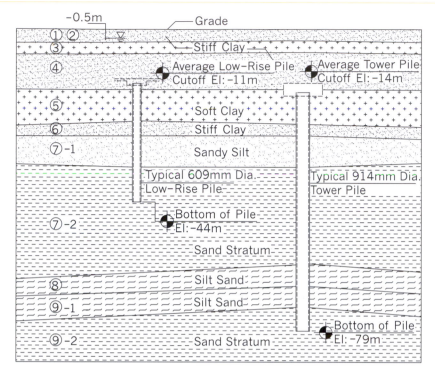

Soil Strata/Extent of Pile Foundation,
Jin Mao Tower, Shanghai, China

2.3.6   Deep Foundation Considerations

Sites that do not have reachable bedrock can be considered for these structures; however, foundation systems become increasingly complex with both strength and settlement issues being critical. A bearing capacity of 480 kPa (10 ksf) usually represents a minimum threshold for design. A bearing capacity of 1900–2400 kPa (40–50 ksf) is more desirable. Pile or caisson foundations allow for adequate support where both skin friction and tip bearing can be used for the design. Piles or caissons should extend 3.0–4.5 m (10–15 ft) into bedrock through a top plane of weathered material that usually exists. Where bedrock does not exist, piles or caissons can be supported in deep stiff sands or hardpan clays. Care should be taken in establishing bearing elevations. Strength may be satisfied at certain soil layers, but these layers may exist over lower compressible layers that could cause adverse long-term settlement. Settlements of 75–125 mm (3–5 in) are not uncommon for pile supported (driven steel, precast concrete, or auger-cast concrete) ultra-tall structures. These settlements must be carefully considered for buildings with entrance levels at grade or interfaces with neighboring structures such as pedestrian tunnels.

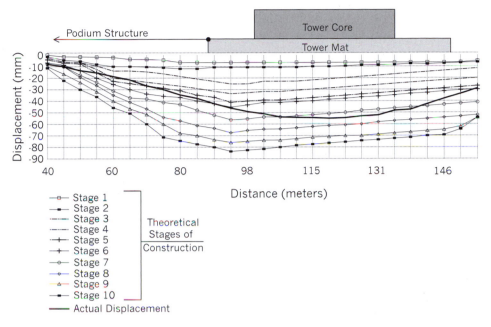

Foundation Settlements,
Jin Mao Tower, Shanghai, China

Differential settlements of foundations are far more serious. Elastic shortening of steel/precast piles and creep, shrinkage, and elastic shortening of cast-in-place piles or caissons must be considered. Uneven lengths of pile or caisson foundations require consideration for applied stress and the length subjected to sustained loads. Longer piles may need to have their cross-section oversized to control this behavior. Special site conditions during construction also must be considered. Pressure grouting of soil for stabilization or for control of ground water infiltration could result in uneven subgrade moduli. Until load is evenly distributed or forces in piles are mobilized through fracturing areas of grouting, towers may settle unevenly which could result in a serious out-of-plumb condition causing global overturning bending moments due to eccentrically placed gravity loads.

CHAPTER 3
FORCES

3.1   CODE-DEFINED GRAVITY LOADS

Beyond self-weight of the structure (based on density of material used) several superimposed types of dead and live loads must be considered in design. The superimposed dead loads are attributed to partitions, ceilings, mechanical systems, floor finishes etc., while superimposed live loads are attributed to occupancies which may vary from residential to office to retail etc. The following is a list of typically anticipated superimposed dead and live loads as defined by codes.

Superimposed dead load (SDL):
- Partitions (dry wall)   = 1.0 kPa (20 psf)
- Ceiling (panel system)   = 0.15 kPa (3 psf)
- Mechanical systems   = 0.10 kPa (2 psf)
- Library or storage   = 7.5 kPa (150 psf)

Live loads (LL):
- Office   = 2.5 kPa (50 psf)
- Residential/hotel   = 2.0 kPa (40 psf)
- Public spaces (i.e. lobbies) = 5.0 kPa (100 psf)
- Parking (passenger veh.)   = 2.0 kPa (40 psf)

3.2   CODE-DEFINED VERTICAL FORCE DISTRIBUTION FOR WIND

Wind loads as defined by the basic equation in Chapter 2.1 are the basis for wind load magnitude. Wind forces generally affect the windward (in direct path of wind) and leeward (opposite face) sides of the structure. Wind loads vary with height, increasing with distance from the ground. These loads must be

FACING PAGE
Poly International Plaza,
Guangzhou, China

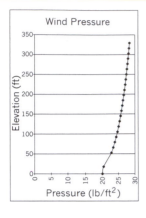
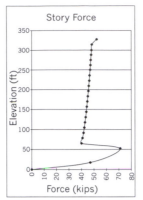

Applied Wind Pressures and Resulting Tower Story Forces

applied to the face area of the structure and consider both windward and lee-ward effects. The image considers a wind force distribution along the height of the tower given the site design criteria. Also included is the force distribution (based on tributary width) and resulting cumulative shears within the tower.

3.3   CODE-DEFINED VERTICAL FORCE DISTRIBUTION FOR SEISMIC

Seismic loads as defined by the 2006 International Building Code (IBC) and the 1997 Uniform Building Code (UBC) are calculated as follows given the site conditions defined.

3.3.1   Earthquake Force

3.3.1.1   Earthquake Force (E)—IBC 2006
The following is the general definition of the earthquake force (E) that must be considered to act on a structure:

$$E = E_h \pm E_v$$

where,

E_h = earthquake force due to the base shear (V) = ρQ_E
E_v = $0.2 S_{DS} D$ = the load effect resulting from the vertical compo-nent of the earthquake motion
S_{DS} = design spectral response acceleration at short periods
D = effect of dead load
E_h = ρQ_E

ρ = redundancy factor

Q_E = effects of horizontal forces from V; where required these forces act simultaneously in two directions at right angles to each other

When design requires the inclusion of an overstrength factor, E shall be defined as follows:

$$E_m = E_{mh} \pm E_v$$

where,

$$E_{mh} = \Omega_o Q_E$$

where,

E_m = seismic load effect including overstrength factor

E_{mh} = effect of horizontal seismic forces including structural overstrength

Ω_o = seismic force amplification factor (overstrength factor)

3.3.1.2   Earthquake Force (E)—UBC 1997

The following is the general definition of the earthquake force (E) that must be considered to act on a structure:

$$E = \rho E_h + E_v$$

where,

E_h = earthquake force due to the base shear (V)

E_v = the load effect resulting from the vertical component of the earthquake motion = $0.5 C_a ID$

ρ = reliability/redundancy factor = $2 - \dfrac{20}{\left(r_{max} \sqrt{A_B}\right)}$

$1.0 \le p \le 1.25$ for special moment-resisting frames ≤ 1.5 for other systems

r_{max} = maximum element–story shear ratio. For initial calculations, this is the ratio of shear in each primary load-resisting element. For more refined analyses, this is the ratio of the design story shear in the most heavily loaded single element divided by the total design story shear.

A_B = area at base of building in sq ft

For critical structural system elements expected to remain essentially elastic during the design ground motion to ensure system integrity:

$$E_m = \Omega_o E_h$$

where,

Ω_o = seismic force amplification factor (overstrength factor)

3.3.2 Static Force Procedure

3.3.2.1 Seismic Base Shear (V)—IBC 2006

The following is a static force procedure based on an approximate method for determining base seismic shear considering the design basis ground motion:

$$V = C_s W$$

where,

V = seismic base shear
C_s = the seismic response coefficient
W = the effective seismic weight
$C_s = \dfrac{S_{DS}}{\left(\dfrac{R}{I}\right)}$

where,

S_{DS} = the design spectral response acceleration parameter in the short period range

$$S_{DS} = \frac{2}{3} S_{MS}$$

where,

S_{MS} = the maximum considered earthquake spectral response accelerations for short period

where,

$$S_{MS} = F_a S_s$$

where,

F_a = short period site coefficient
S_s = the mapped maximum considered spectral accelerations for short periods
R = the response modification factor (structural system dependent)
I = occupancy importance factor

However, the value of C_s need not exceed the following:

$$C_s = \frac{S_{D1}}{T\left(\dfrac{R}{I}\right)} \quad \text{for } T \le T_L$$

$$C_s = \frac{S_{D1}T_L}{T^2\left(\dfrac{R}{I}\right)} \quad \text{for } T > T_L$$

C_s shall not be less than: $C_s = 0.044\, S_{DS}\, I \ge 0.01$

In addition, for structures located where S_1 is equal to or greater than 0.6g, C_s shall not be less than:

$$C_s = \frac{0.5S_1}{\left(\dfrac{R}{I}\right)}$$

where,

S_{D1} = the design spectral response acceleration parameter at a period of 1.0s

$$S_{D1} = \frac{2}{3}\, S_{M1}$$

where,

S_{M1} = the maximum considered earthquake spectral response accelerations for a 1-second period

where,

$$S_{M1} = F_v S_1$$

where,

F_v = long period site coefficient

S_1 = the mapped maximum considered spectral accelerations for a 1-second period

T = the fundamental period of the structure

T_L = long period transition period(s)

$T = C_t(h_n)^x$

where,

C_t = 0.028,   x = 0.8 for steel moment-resisting frames

C_t = 0.016,   x = 0.9 for concrete moment-resisting frames

C_t = 0.03,   x = 0.75 for steel eccentrically braced frames

C_t = 0.02,   x = 0.75 for all other structural systems

h_n = height from the base of the building to the highest level (feet)

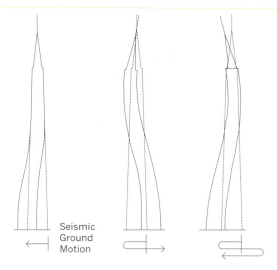

Tower Motion Resulting from Seismic Ground Motion

Alternatively, it is permitted to determine the approximate fundamental period (T_a) in seconds (s) from the following equation for structures not exceeding 12 stories in height in which the seismic force-resisting system consists entirely of concrete or steel moment-resisting frames and the story height is at least 3 m (10 ft):

$$T_a = 0.1N$$

where,

$\qquad N$ = number of stories

The approximate fundamental period, T_a, for masonry or concrete shear wall structures is permitted to be determined as follows:

$$T_a = \frac{0.0019h_n}{\sqrt{C_W}}$$

where,

$$C_W = \frac{100}{A_B} \sum_{i=1}^{x} \left(\frac{h_n}{h_i}\right)^2 \frac{A_i}{1 + 0.83\left(\frac{h_i}{D_i}\right)^2}$$

where,

$\qquad A_B$ = area of base of structure in sq ft
$\qquad A_i$ = web area of shear wall "i" in sq ft

D_i = length of shear wall "i" in ft
h_i = height of shear wall "i" in ft
x = number of shear walls in the building effective in resisting lateral forces in the direction under consideration

3.3.2.2 Seismic Dead Load (W)—IBC 2006

Applicable portions of other loads to be considered for the total seismic load, W, include:

1. In storage and warehouse occupancies, a minimum of 25% of the floor live load.
2. Where partition loads are used in floor design, a load not less than 10 psf.
3. Total operating weight of permanent equipment.
4. Where the flat roof snow load exceeds 30 psf or 20% of the uniform design snow load regardless of actual roof slope.

3.3.2.3 Seismic Base Shear (V)—UBC 1997

The following is a static force procedure based on an approximate method for determining base shear considering the design basis ground motion:

$$\frac{2.5C_a I}{R} W \geq V = \frac{C_v I}{RT} W \geq 0.11 C_a I W$$

$$\geq \frac{0.8 Z N_v I}{R} W \text{ (Seismic Zone 4)}$$

where,

C_a and C_v = seismic coefficients
I = seismic importance factor
W = total seismic dead load (total dead load plus applicable portions of other loads)
R = response modification factor
T = fundamental period of vibration of the structure
Z = seismic zone factor
N_v = velocity-dependent near-source factor

3.3.2.4 Seismic Dead Load (W)—UBC 1997

Applicable portions of other loads to be considered for the total seismic load, W, include:

1. In storage and warehouse occupancies, a minimum of 25% of the floor live load.
2. Where partition loads are used in floor design, a load not less than 10 psf.
3. 20% of the uniform design snow load, when it exceeds 30 psf.
4. Total weight of permanent equipment.

3.3.2.5 Fundamental Period (Approximate Methods)—UBC 1997

For the determination of building period (T), by the approximate Method A:

$$T = C_t(h_n)^{3/4}$$

where,

C_t = 0.035 for steel moment-resisting frames

C_t = 0.030 for reinforced concrete moment-resisting frames and eccentrically braced frames

C_t = 0.020 for other concrete buildings

where,

h_n = height from the base of the building to the highest level (ft)

Alternatively, for structures with concrete or masonry shear walls:

$$C_t = \frac{0.1}{\sqrt{A_c}}$$

where,

A_c = combined effective area of shear walls in the first story of the structure (sq ft) =

$$\sum A_e\left[0.2 + (D_e/h_n)^2\right], \ \frac{D_e}{h_n} \leq 0.9$$

A_e = minimum cross-sectional area in any horizontal plane in the first story of a shear wall (sq ft)

D_e = length of shear wall in the first story in the direction parallel to the applied forces (ft)

Once preliminary sizes are obtained based on the base shear calculated using the approximate period T, a more accurate value of T can be determined using established analytical procedures.

In lieu of approximate Method A, Method B provided in the UBC Code can be used to determine T. Method B permits the evaluation of T by either the Rayleigh formula or other substantiated analysis. Note that the value of T obtained from Method B must be less than or equal to 1.3 times the value of T obtained from Method A in Seismic Zone 4, and less than or equal to 1.4 times the value in Seismic Zones 1, 2, and 3.

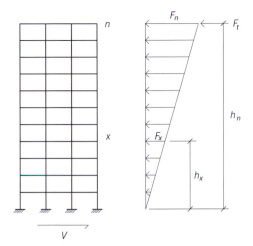

Vertical Force Distribution for Seismic Loading

3.3.3 Distribution of Lateral Forces

3.3.3.1 Vertical Force Distribution
The base shear (V) is distributed vertically to each floor level of the building. The story shears are then distributed to the lateral force-resisting elements proportional to their relative stiffness and the stiffness of the diaphragms.

As described in the UBC Code, the base shear is distributed linearly over the height of the building, varying from zero at the bottom to a maximum at the top, corresponding to the fundamental (first mode) period of vibration of the structure. To account for higher mode effects (buildings with a fundamental period greater than 0.7 seconds) on the structure, a portion of the base shear is applied as a concentrated load at the top of the building (see Section 3.3.3.2).

3.3.3.2 Horizontal Force Distribution
The seismic design story shear in any story (V_x) is determined as follows:

$$V_x = F_t + \sum_{i=x}^{n} F_i$$

and the base shear is:

$$V = F_t + \sum_{i=1}^{n} F_i$$

where,

$$F_t = 0.07TV < 0.25V \text{ for } T > 0.7 \text{ seconds}$$
$$= 0.0 \text{ for } T \le 0.7 \text{ seconds}$$

and,

$$F_x = \frac{(V - F_t)w_x h_x}{\displaystyle\sum_{i=1}^{n} w_i h_i}$$

where,

$F_i F_n F_x$ = design seismic force applied to level i, n, or x, respectively

F_t = portion of V considered concentrated at the top of the structure in addition to F_n

$h_i h_x$ = height above the base to level i or x, respectively

$w_i w_x$ = the portion of W located at or assigned to level i or x, respectively

T = fundamental period of vibration of structure in seconds in direction of analysis

3.3.4   Bending Moment Distribution (Overturning)

Once design seismic forces applied to levels have been established, the bending moment due to these forces can be determined. The tower structure must be designed to resist the overturning effects caused by the earthquake forces. The overturning moment (M_x) at any level x can be determined by the following formula:

$$M_x = \sum_{i=x}^{n} F_i(h_i - h_x) + F_t(h_n - h_x)$$

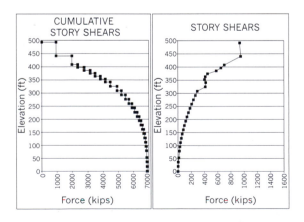

Cumulative and Individual Story Shears Resulting from Seismic Loading

where,

F_i = portion of seismic base shear (V) located or assigned to level i

F_t = portion of V considered concentrated at the top of the structure in addition to F_n

h_i, h_n, h_x = height above the base to level i, n, or x, respectively

3.3.5   Story Drift Limitations

Drift is defined as the displacement that a structure experiences when subjected to load. Drift is usually caused by lateral loads due to wind or seismic events but could be caused by unbalanced gravity loads or temperature effects disproportionately applied. The structure experiences overall drift, which is described as the displacement at the top of the building relative to the ground.

Inter-story drift is the relative displacement of one floor level to another. For seismic events this calculation is important because inter-story drifts due to inelastic response could be large. Exterior wall and partition systems among other vertical building systems must be detailed to allow for this movement.

Inter-story drifts within the structure shall be limited to a maximum inelastic drift response, approximately equal to the displacement that occurs in the structure when subjected to the design basis ground motion:

$$\Delta_M = 0.7 R \Delta_s$$

where,

Δ_M = maximum expected inelastic drift

R = response modification factor

Δ_s = maximum computed elastic drift considering the lateral force-resisting system

For structures with a fundamental period (T) less than 0.7 seconds, the calculated story drift using Δ_M shall not exceed 0.025 (2.5%) times the story height. For structures with T greater than or equal to 0.7 seconds, the story drift shall not exceed 0.02 (2%) times the story height.

3.4   GRAVITY LOAD DISTRIBUTION AND TAKEDOWNS

3.4.1   Floor Systems

Gravity loads are generally considered to be uniformly distributed over an occupied floor. These loads vary based on building use and include dead load (self-weight), superimposed dead load (load from building components that

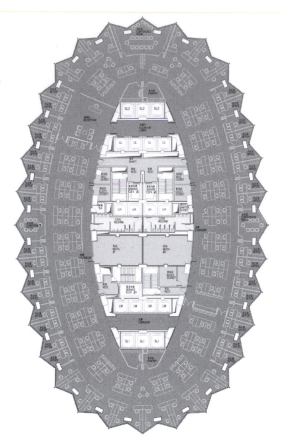

Proposed Occupancy Use of a Typical Floor Plan,
Jinta Tower, Tianjin, China

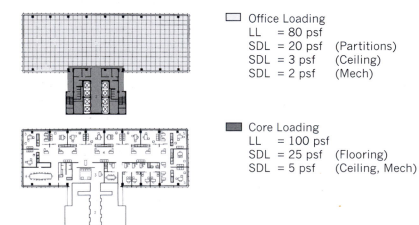

Office Loading
LL   = 80 psf
SDL = 20 psf   (Partitions)
SDL = 3 psf   (Ceiling)
SDL = 2 psf   (Mech)

Core Loading
LL   = 100 psf
SDL = 25 psf   (Flooring)
SDL = 5 psf   (Ceiling, Mech)

Design Floor Loading, Typical Office Floor Plan

have little variation in magnitude of load over the life of the building, i.e. partitions, ceiling systems, mechanical systems), and live load (which can vary in magnitude and location).

When considering dead load (self-weight), all components of the primary structure must be included, typically floor slabs, floor framing beams and girders, and columns. Material density must be known in order to accurately calculate self-weight of the structure. Common densities include structural steel 7850 kg/cu m (490 lbs/cu ft) and reinforced concrete 2400 kg/cu m (150 lbs/cu ft).

Common loads in a building structure are listed as follows:

Superimposed dead load (SDL):

Partitions (dry wall)	= 1.0 kPa (20 psf)
Ceiling (panel system)	= 0.15 kPa (3 psf)
Mechanical systems	= 0.10 kPa (2 psf)
Library or storage	= 7.5 kPa (150 psf)
Finished flooring	= 1.2 kPa (25 psf)

Live loads (LL):

Office	= 2.5 kPa (50 psf)
Office (premium)	= 4.0 kPa (80 psf)
Residential/hotel	= 2.0 kPa (40 psf)
Public spaces (i.e. lobbies)	= 5.0 kPa (100 psf)
Parking (passenger veh.)	= 2.0 kPa (40 psf)

There are cases where individual concentrated point or specific line loads must be considered on the structure. An example of a concentrated point load may include the consideration of truck loadings in a loading dock, and an example of a specific line load may include the consideration of a heavy masonry partition used for acoustic isolation.

3.4.2   Exterior Walls

Exterior walls for tower structures produce specific loads that must be considered in the structure design. These loads may be light (e.g. metal panels and glass) or heavy (e.g. precast concrete) depending on the architectural design. Based on exterior wall connections, imposed loads can be calculated. In some cases, the exterior wall is supported on perimeter girders and in some cases attached directly to columns. For initial calculations, exterior wall loads may be considered to be evenly distributed along perimeter spandrels, considering the exterior wall weight and the floor-to-floor heights.

Exterior wall loads are often considered as distributed load over the "face" area of the structure. Some common exterior wall loads are as follows (all of which would need to be confirmed based on the final as-designed exterior wall system):

> Metal and glass   = 0.75 kPa (15 psf)
> Stone and glass   = 1.2 kPa (25 psf)
> Precast and glass = 2.5 kPa (50 psf)

3.4.3 Loads to Vertical Elements

Loads, either distributed or concentrated, are first typically supported by horizontal framing members, then by vertical columns or walls, then by foundation systems.

Loads distributed to floor slabs are typically supported by beam framing, then girder framing, then by columns or walls. Knowing the required spans and support conditions of the floor framing elements, distributed loads are used for the design of the members. These loads then transfer through the horizontal support systems to vertical load carrying elements. Generally, columns and walls support a tributary floor area, as well as a tributary exterior wall area with the deletion of any floor or wall openings. A column or wall load takedown is performed on each discrete column or wall element from the influence of multiple floors on the elements. Codes typically recognize a reduction in live load on vertical elements when multiple floors and large areas are considered.

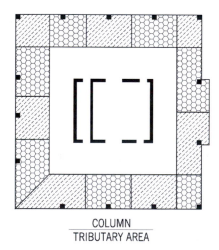

COLUMN
TRIBUTARY AREA

Tributary Loading Area to Columns

3.5   LOAD COMBINATIONS

In accordance with IBC 2006 the following load combinations shall be used.

3.5.1   Basic Load Combinations —Strength or Load and Resistance Factor Design

1. $1.4(D + F)$
2. $1.2(D + F + T) + 1.6(L + H) + 0.5(L_r \text{ or } S \text{ or } R)$
3. $1.2D + 1.6(L_r \text{ or } S \text{ or } R) + (f_1 L \text{ or } 0.8W)$
4. $1.2D + 1.6W + f_1 L + 0.5(L_r \text{ or } S \text{ or } R)$
5. $1.2D + 1.0E + f_1 L + f_2 S$ or
   $(1.2 + 0.2S_{DS})D + \rho Q_E + L + 0.2S$ or
   $(1.2 + 0.2S_{DS})D + \Omega_o Q_E + L + 0.2S$—when considering structural overstrength
6. $0.9D + 1.6W + 1.6H$
7. $0.9D + 1.0E + 1.6H$ or $(0.9 - 0.2S_{DS})D + \rho Q_E + 1.6H$ or
   $(0.9 - 0.2S_{DS})D + \Omega_o Q_E + 1.6H$—when considering structural overstrength

where,

f_1 = 1 for floors in places of public assembly, for live loads in excess of 5 kPa (100 psf), and for parking garage live load, and = 0.5 for other live loads

f_2 = 0.7 for roof configurations (such as saw tooth) that do not shed snow off the structure, and = 0.2 for other roof configurations

Exception: Where other factored load combinations are specifically required by the provisions of the IBC 2006 Code, such combinations shall take precedence.

3.5.2   Basic Load Combinations—Allowable (Working) Stress Design

1. $D + F$
2. $D + H + F + L + T$
3. $D + H + F + (L_r \text{ or } S \text{ or } R)$
4. $D + H + F + 0.75(L + T) + 0.75(L_r \text{ or } S \text{ or } R)$
5. $D + H + F + (W \text{ or } 0.7E)$ or
   $(1.0 + 0.14S_{DS})D + H + F + 0.7\rho Q_E$ or
   $(1.0 + 0.14S_{DS})D + H + F + 0.7\Omega_o Q_E$
   when considering structural overstrength

6. $D + H + F + 0.75(W \text{ or } 0.7E) + 0.75L + 0.75(L_r \text{ or } S \text{ or } R)$ or
$(1.0 + 0.105S_{DS})D + H + F + 0.525\rho Q_E + 0.75L + 0.75(L_r \text{ or } S \text{ or } R)$
or
$(1.0 + 0.105S_{DS})D + H + F + 0.525\Omega_o Q_E + 0.75L + 0.75(L_r \text{ or } S \text{ or } R)$
when considering structural overstrength
7. $0.6D + W + H$
8. $0.6D + 0.7E + H$ or $(0.6 - 0.14S_{DS})D + 0.7\rho Q_E + H$ or
$(0.6 - 0.14S_{DS})D + 0.7\Omega_o Q_E + H$
when considering structural overstrength

where,

D = dead load
E = combined effect of horizontal and vertical earthquake induced forces as defined in Section 3.3 (or Section 12.4.2 of ASCE 7)
E_m = maximum seismic load effect of horizontal and vertical seismic forces as set forth in Section 3.3 (or Section 12.4.3 of ASCE 7)
F = load due to fluids with well-defined pressures and maximum heights
H = load due to lateral earth pressure, ground water pressure or pressure of bulk materials
L = live load, except roof live load, including any permitted live load reduction
L_r = roof live load including any permitted live load reduction
R = rain load
S = snow load
T = self-straining force arising from contraction or expansion as a result of temperature change, shrinkage, moisture change, creep in component materials, movement due to differential settlement or combinations thereof
W = load due to wind pressure

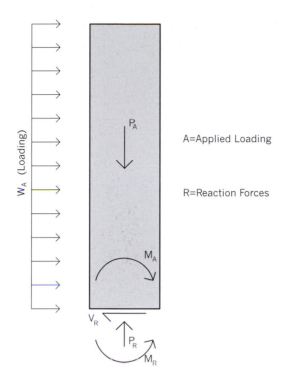

A=Applied Loading

R=Reaction Forces

Applied Load and Resulting Forces/Reactions on Tower

3.6   DESIGN AXIAL, SHEAR, AND BENDING MOMENTS

Once the gravity and lateral loads along with the controlling load combinations have been established, the design axial, shear, and bending moments both globally and on individual structural elements can be established. These loads will be used in the design of the structure based on material types and behavior.

CHAPTER 4
LANGUAGE

4.1   FORCE FLOW

ONE OF THE MOST IMPORTANT, and likely the most obscure, considerations is reading the force flow through a structure. An accurate understanding of this flow leads to the most correct assessment of behavior and the safest and most efficient design. These forces primarily originate from gravity loads and lateral loads caused by wind and seismic events. Other forces may be caused by settlement, temperature or relative displacements of vertical columns or walls due to creep, shrinkage, and elastic shortening, etc. Once the loads acting on the structure are fundamentally understood, the flow of these forces must be understood. Forces typically flow through floor framing systems into vertical elements such as columns or walls into foundations. Lateral force-resisting systems are typically subjected to temporary loads, externally imposed, with flow through vertical systems that are supported by foundation systems. Lateral force-resisting systems often resist gravity loads in addition to temporarily imposed lateral loads. These gravity loads, when strategically placed, actually can work to an advantage in the structure,

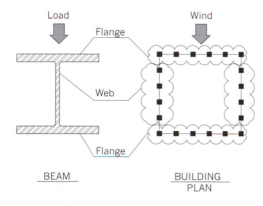

BEAM   BUILDING   Wide-Flanged Beam Analogy
                        PLAN   to Tower Plan

FACING PAGE
John Hancock Center,
Chicago, IL

53

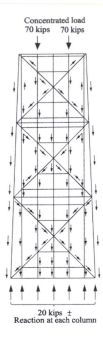

Force Flow,
John Hancock Center, Chicago, IL

acting as a counterbalance to overturning effects and applying "prestress" to members that would otherwise be subjected to tension when subjected to lateral loads.

4.2   STRUCTURAL FRAMING PLANS

4.2.1   Lateral vs. Gravity Systems

Lateral and gravity system components often are not defined specifically in a framing plan. These framing plans typically only show a portion of the structure with vertical components generally indicated but more completely described in overall structural system elevations and sections.

In the definition of force flow, the evaluation usually begins in plan, with elevations and sections following, and foundation systems last. Forces superimposed on floors migrate from elements with least stiffness to those with greatest stiffness. The typical migration originates in slab framing to floor beams or trusses, to floor girders, to columns or walls. This behavior is the same for all materials; however, in some structures floor beams or girders may not exist with systems only incorporating flat slabs or "plates."

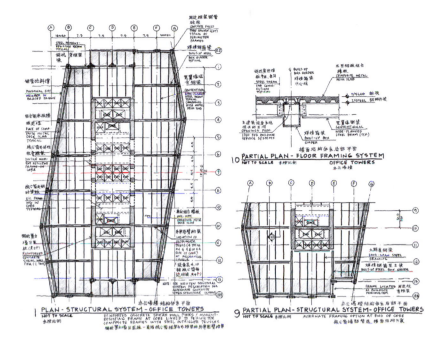

Tower Structural System Plans and Detail,
Hang Lung Competition, China

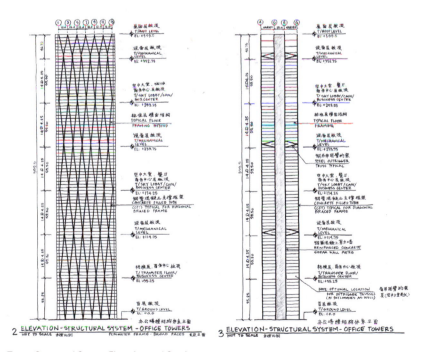

Tower Structural System Elevation and Section,
Hang Lung Competition, China

4.2.2 Steel

John Hancock Center, Chicago, Illinois—Structural steel floor framing spans from the interior structural steel columns resisting gravity loads only (typical connections from beam framing to interior columns consist of shear web connections only without flange connections) to the exterior braced frame where connections of diagonal, vertical, and horizontal members develop the full moment capacity of the members. The exterior tubular braced frame resists gravity and lateral loads.

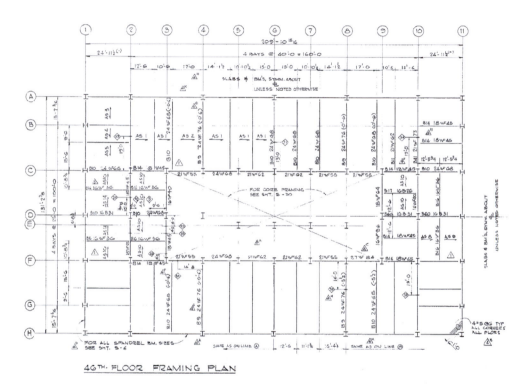

Floor Framing Plan,
John Hancock Center, Chicago, IL

222 South Main Street, Salt Lake City, Utah—Structural steel floor framing spans from the interior concentrically braced frames (unbonded braces) to perimeter moment-resisting frames (beam-to-column connections are fully welded to develop full moment and shear capacity of the joints).

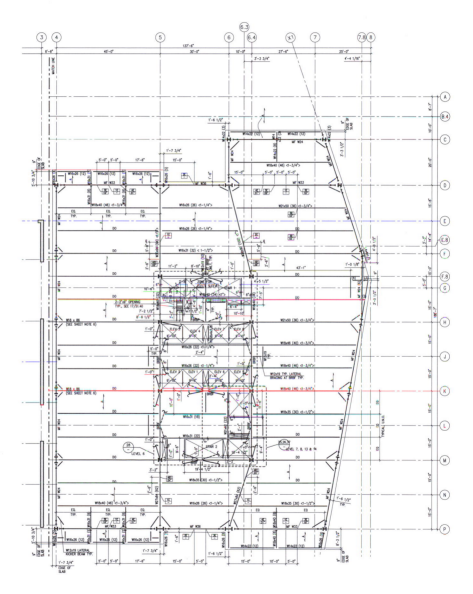

Floor Framing Plan,
222 South Main Street, Salt Lake City, UT

4.2.3   Concrete

500 West Monroe, Chicago, Illinois—Conventional long-span reinforced concrete framing spans from the perimeter reinforced concrete frames to the central reinforced concrete shear wall core. The perimeter frames and the central core resist both lateral and gravity loads.

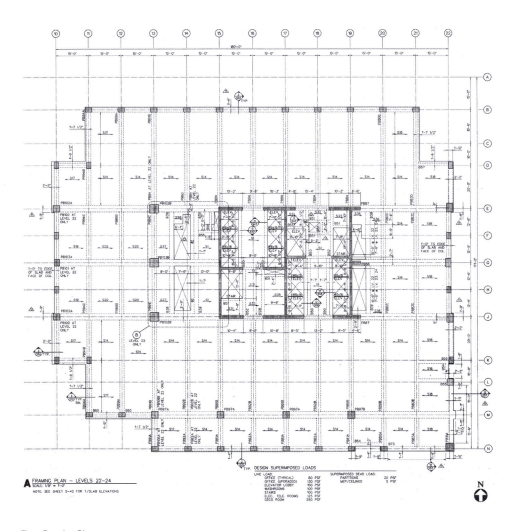

Floor Framing Plan,
500 West Monroe, Chicago, IL

University of California, Merced, Kolligian Library, Merced, California—Conventional long-span reinforced concrete framing spans from internal reinforced concrete frames in each primary building direction.

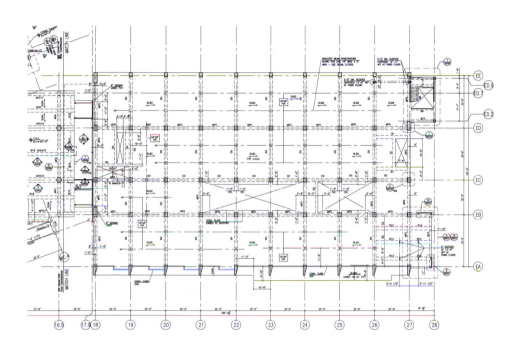

Floor Framing Plan,
UC Merced Kolligian Library, Merced, CA

4.2.4   Composite

New Beijing Poly Plaza, Beijing, China—Structural steel floor framing spans from central reinforced concrete shear wall cores to perimeter steel moment-resisting frames. End connections for typical framing members use bolted shear tab connections with perimeter frame connections using a combination of bolts and welds to fully develop moment capacity of frame members. The shear wall cores and perimeter frames resist both gravity and lateral loads.

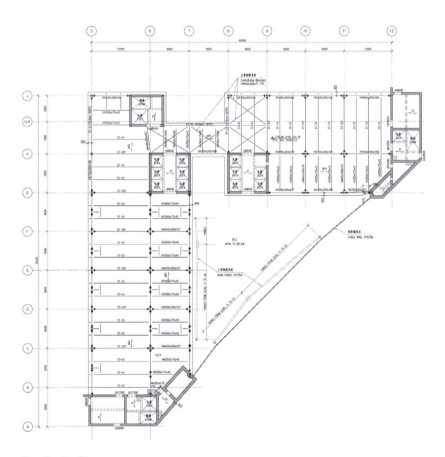

Floor Framing Plan,
New Beijing Poly Plaza, Beijing, China

Jinao Tower, Nanjing, China—Reinforced concrete long-span conventional framing spans between central reinforced concrete shear wall core and perimeter reinforced concrete tubular frame. Diagonal structural steel tubes are located on the outside of the perimeter tube to provide increased lateral load resistance. The central core, perimeter frame, and the diagonal braces all provide lateral load resistance, while the core and perimeter frame also resist gravity loads.

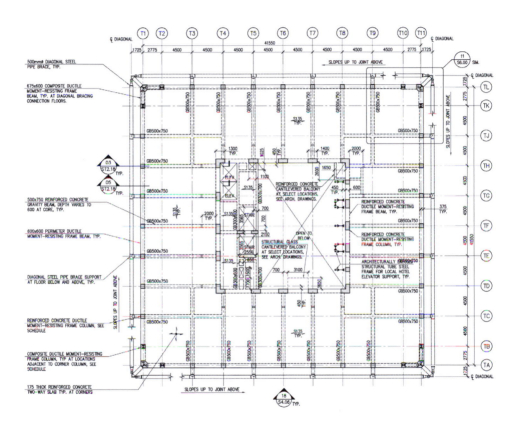

Floor Framing Plan,
Jinao Tower, Nanjing, China

4.3   STRUCTURAL SYSTEM ELEVATIONS

4.3.1   Steel

Willis Tower (formerly Sears Tower), Chicago, Illinois—Bundled steel tubular frame is used to resist lateral and gravity loads. Frames exist at the perimeter and in internal locations. Steel belt trusses are used to transfer lateral loads when tubular frame steps in elevation.

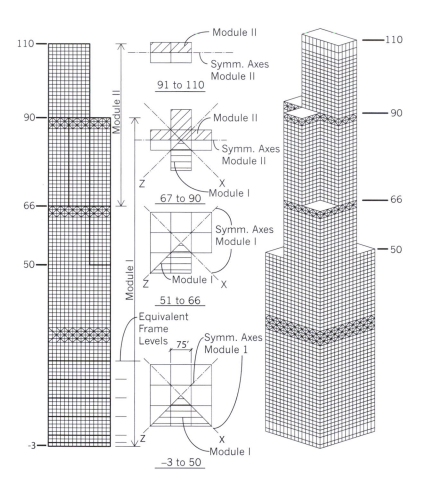

Lateral System Elevation,
Willis Tower (formerly Sears Tower), Chicago, IL

Tustin Legacy Park, Tustin, California—Concentrically and eccentrically braced shear truss core combined with a perimeter steel frame resist lateral and gravity loads.

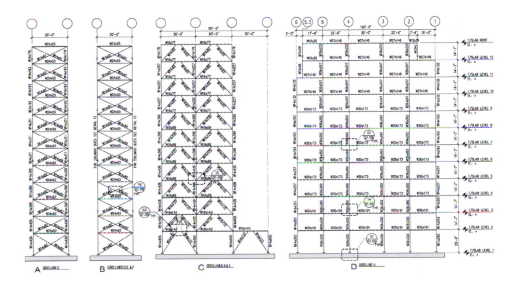

Lateral System Elevation,
Tustin Legacy Park, Tustin, CA

4.3.2   Concrete

Goldfield International Garden, Beijing, China—Reinforced concrete superframe infilled with irregular screen frames on two of four facades and conventional reinforced concrete moment-resisting frames on the other two facades plus the central reinforced concrete shear wall core resist lateral and gravity loads.

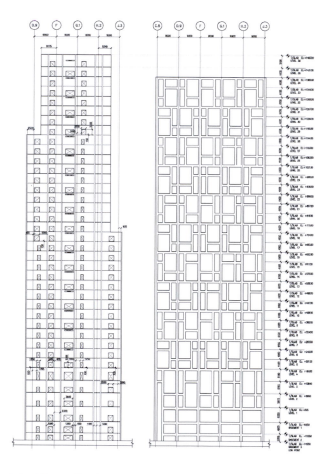

Lateral System Elevation,
Goldfield International Garden, Beijing, China

Burj Khalifa, Dubai, UAE—Central reinforced concrete shear wall buttressed core interconnected to perimeter reinforced concrete mega-columns resist both lateral and gravity loads.

Construction Image,
Burj Khalifa, Dubai, UAE

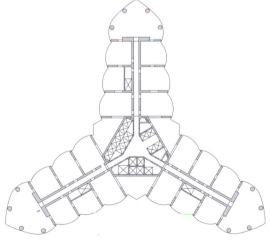

Floor Plan,
Burj Khalifa, Dubai, UAE

4.3.3 Composite

Jin Mao Tower, Shanghai, China—A central reinforced concrete shear wall core is interconnected with composite mega-columns through outrigger trusses at three two-story levels. The central core and the perimeter mega-columns resist both gravity and lateral loads.

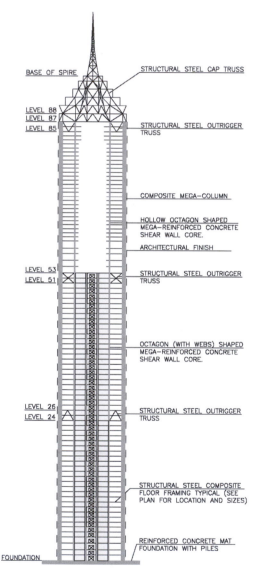

Lateral System Elevation,
Jin Mao Tower, Shanghai, China

Jinta Tower, Tianjin, China—A central steel-plated core (circular composite columns are interconnected by unstiffened steel plates) is interconnected with a perimeter composite moment-resisting frame.

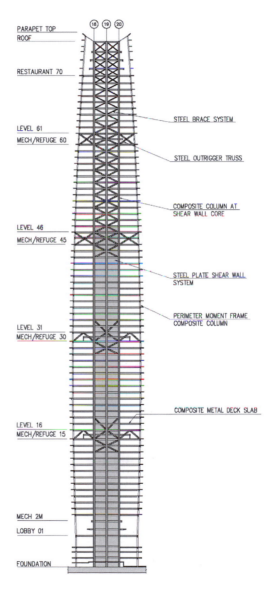

Lateral System Elevation,
Jinta Tower, Tianjin, China

CHAPTER 5
ATTRIBUTES

STRENGTH, INCLUDING CODE LIMITS and material types, and serviceability, including drift, damping, wind-induced accelerations, creep, shrinkage, elastic shortening, have a fundamental effect on the design of the tower. The understanding of materials, building proportioning, and building behavior when subjected to loads are critical in determining both feasibility and successful use.

5.1   STRENGTH

Whether limit state (load and resistance factor design) or allowable stress design is used for members within the structural system, local codes and material type will dictate the method of design. Redundancy, load path, and the importance of structural elements within the system result in special structural design considerations. Strength design is usually based on wind loads with a return period of 50 years and seismicity with a 10% probability of exceedance in 50 years adjusted by a structural system-dependent reduction factor (response modification factor). The historic reliance on structural steel for tall towers has evolved into the use of ultra-high strength concrete with common production compressive strengths of 110 MPa (16,000 psi) and higher, compressive strengths approaching that of early cast irons/ structural steels. Grade 36 steel has merged with Grade 50 steel and higher strength steels with yield strengths of 450 MPa (65 ksi) and higher are commonplace.

This increase in available strengths for concrete and steel has allowed more efficient designs with smaller structural elements. In addition, the combination of structural steel and reinforced concrete (composite) in structures has led to extremely efficient solutions.

FACING PAGE
Burj Khalifa Under Construction,
Dubai, UAE

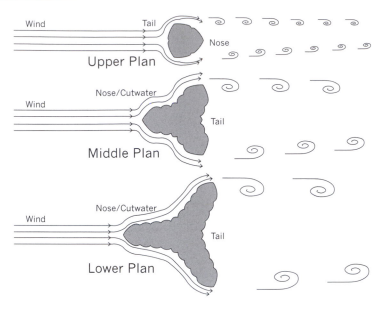

Disruption of Organized Vortex Shedding,
Burj Khalifa, Dubai, UAE

5.2   SERVICEABILITY

In addition to strength considerations, serviceability of the tall tower is likely the most important design consideration and, at times, the least understood. Most consider drift to be the controlling factor for stiffness; in fact, building acceleration due to wind-induced motion can be far more critical. Evaluating occupants' perception of motion is based on building use, stiffness, mass, and damping of the structure.

5.2.1   Drift

The internationally recognized drift criterion for ultra-tall structures is h/500, typically based on elastic deformations (cracked sections in some reinforced concrete members including link beams and moment-frame beams) and a 50-year return wind. Historically, some tall structures have been designed with allowable drifts as high as h/400.

In developing rational, applied wind pressure diagrams for the structure, specific damping ratios should be considered based on building materials and non-structural components (see following section for damping).

It is difficult to find codes that commit to the allowable building drifts for structures subjected to wind. The Canadian Building Code (h/500 for all structures) and the Chinese National Building Code are exceptions. On

Deflection Shape Due to Seismic Loading,
Agile Tower, Guangzhou, China

the other hand, seismic drift limits are recognized by the UBC and are dependent on the structural system used. The limit is as follows:

$$\Delta m = 0.7 \, R \, \Delta s$$

where,

Δm = maximum inelastic response drift (Δm shall not exceed .025h for T<0.7 seconds and 0.02h for T≥0.7 seconds, or h/40 and h/50 respectively)

R = lateral system coefficient representative of inherent over-strength and global ductility

Δs = maximum elastic drift

Some codes, such as China's National Building Code, take a strict position on allowable building drift for both wind and seismic conditions with system specific criteria. The limits are based on a 50-year return wind, frequent earthquakes (62.5% probability of exceedance in 50 years), and elastic section properties (including gross section properties for reinforced concrete). The following chart summarizes major systems and limits.

Structural system	Any height	H < 150M	H ≥ 150M H ≤ 250M	H > 250M
Steel				
Wind (roof)	h/500	–	–	–
Wind (interstory)	h/400	–	–	–
Seismic (frequent)	h/250	–	–	–
Seismic (rare)	h/70	–	–	–
Concrete (wind and seismic frequent)				
Frame	–	h/550	Interpolate	h/500
Frame—shear wall (SW)	–	h/800	Interpolate	h/500
SW only, tube-in-tube	–	h/1000	Interpolate	h/500
Composite (wind and seismic frequent)				
Steel frame—SW	–	h/800	Interpolate	h/500
Comp frame—SW	–	h/800	Interpolate	h/500
Comp frame (steel beams)	h/400	–	–	–
Comp frame (comp beams)	h/500	–	–	–
Concrete frame (seismic rare)				
Frame	h/50	–	–	–
Frame—SW	h/100	–	–	–
SW—tube-in-tube	h/120	–	–	–
Composite (seismic rare)				
Comp frame	h/50	–	–	–
Other	h/100	–	–	–

Second order (P-Δ) effects must be considered in tall structures laterally displaced by wind and seismic loads. These effects could increase drift by as much as 10% and must be also considered for the strength design of lateral load-resisting members. Examples of building drifts for major building projects are listed as follows:

Building	Height	Drift	Material
Willis Tower (Chicago) (formerly Sears Tower)	445 m	H/550	Steel
Jin Mao Tower (Shanghai)	421 m	H/908	Mixed
Central Plaza (Hong Kong)	374 m	H/780	Concrete
Amoco Building (Chicago)	346 m	H/400	Steel
John Hancock (Chicago)	344 m	H/500	Steel
Columbia Seafirst (Seattle)	288 m	H/600	Mixed
Citibank Plaza (Hong Kong)	220 m	H/600	Mixed

5.2.2   Damping

Damping of the tall towers can have a significant effect on design forces and wind-induced accelerations. Damping is material specific and is proportional to demand on the structure. Theoretical analysis, laboratory testing, and in-situ monitoring have provided general requirements for design. The overall building damping typically considered in the behavior of tall building structures is as follows (% of critical):

Material	Return period for wind			
	1–10 year	50 year	100 year	1000 year (collapse prev.)
Concrete	2%	3%	5%	7%
Steel	1%	2%	3%	4%

The building damping values may be specifically calculated as follows:

$$\xi = \xi_N + \xi_M + \xi_{SD} + \xi_{AE} + \xi_{SDS}$$

where,

ξ = total building damping ratio

ξ_N = non-structural component damping (1–1.5%)

ξ_M = material damping (concrete uncracked members = 0.75%, steel = 0%)

ξ_{SD} = structural damping (concrete cracked members = 0.5–1.5%, steel = 0–0.5%)

ξ_{AE} = aero-elastic damping (0–0.75%)

ξ_{SDS} = supplemental damping systems (visco-elastic = 5–30%, vibration absorbers = 1–5%)

and,

$$\xi_{SD} = \frac{\sum E_D}{4\pi E_{SO}}$$

where,

E_D = energy loss per element per full cycle to a given performance level

E_{SO} = total building elastic strain energy associated with a given performance level

$$E_{SO} = \sum \tfrac{1}{2} F_i \Delta_i$$

where,

F_i = wind force at each level for a given direction

Δ_i = corresponding displacement at the point of loading, at each level

$$\Sigma E_D = k^* \sum_i E_{SD}$$

where,

k = adjustment factor, $\quad k = \dfrac{4\pi(E_{SOmodel})\xi_{SD\,"measured"}}{\sum_i E_{SDmodel}}$

E_D = total energy loss in the members ("measured")

E_{SD} = total strain energy in the members (model)

5.2.3 Accelerations

Without acceptable levels of wind-induced accelerations, the tall tower can be unusable during strong wind events. Across-wind accelerations or lift accelerations are usually more serious than along-wind or drag accelerations. There have been recorded instances where occupants of super-tall buildings have perceived building motion, felt nauseous, and in some cases left the building during a windstorm. Other visual or audible conditions during windstorms lead to discomfort. Perceived motion relative to neighboring structures is most acute especially due to torsion. Water in toilets may slosh. Exterior wall elements or interior partitions may creak. Wind speed, building height, orientation, shape, and regularity along the elevation all contribute to the behavior.

Apartment in the John Hancock Center, Chicago, IL

The inner ear is very sensitive to motion. A person lying down is more susceptible than one in a seated or standing position. People who reside in structures (apartments or condominiums) rather than occupy them transiently (office buildings) are generally more susceptible to perceiving building movement. The limits of perceptible acceleration are:

	Horizontal accelerations return wind period	
Occupancy type	**1 year**	**10 year**
Office	10–13 milli-g's	20–25 milli-g's
Hotel	7–10 milli-g's	15–20 milli-g's
Apartment	5–7 milli-g's	12–15 milli-g's

These accelerations are usually most critical at the top occupied floor and are calculated for building damping ratios of 1% of critical for steel structural systems to 1½% of critical for composite or reinforced concrete structural systems. The return period refers to the maximum accelerations statistically expected for maximum winds over a defined period of time.

Torsional accelerations/velocities experienced by a structure are in many cases more important than horizontal accelerations. This is especially true where occupants have a point of reference relative to neighboring structures. The limit of acceptable torsional velocity is 3.0 milli-radians/sec. The National Building Code of Canada Structural Commentary Part 4 offers a calculation method to predict horizontal accelerations. This method provides an excellent preliminary calculation that can be later confirmed by the rational wind tunnel studies.

Across-wind accelerations are likely to exceed along-wind accelerations if the building is slender about both axes, that is if:

$$\sqrt{WD}/H < {}^{1}/_{3}$$

where,

W = across-wind plan dimension (m)
D = along-wind plan dimension (m)
H = height of the building (m)

For these tall, slender structures, accelerations caused by wind-induced motion are defined as:

$$a_w = n_w^2 g_p \sqrt{WD} \left(\frac{a_r}{\rho_B g \sqrt{\beta_w}} \right)$$

For less slender structures or for lower wind speeds, the maximum acceleration is:

$$a_D = 4\pi^2 n_D^2 g_p \sqrt{\frac{K_s F}{C_e \beta_D}} \frac{\Delta}{C_g}$$

where,

a_w, a_D = peak acceleration in across-wind and along-wind directions, m/s²

a_r = $78.5 \times 10^{-3} [V_H / (n_w \sqrt{WD})]^{3.3}$, Pa

ρ_B = average density of the building, kg/m³

β_w, β_D = fraction of critical damping in across-wind and along-wind directions

n_w, n_D = fundamental natural frequencies in across-wind and along-wind directions, Hz

Δ = maximum wind-induced lateral deflection at the top of the building along-wind direction, m

g = acceleration due to gravity = 9.81 m/s²

g_p = a statistical peak factor for the loading effect

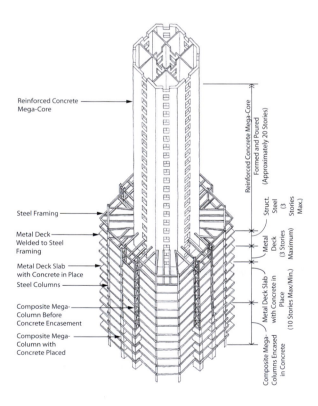

Mixed Use of Structural Materials/Construction Sequence, Jin Mao Tower, Shanghai, China

K = a factor related to the surface roughness coefficient of the terrain

s = size reduction factor

F = gust energy ratio

C_e = exposure factor at the top of the building

C_g = dynamic gust factor

Supplemental damping may be introduced into the structure to control accelerations. This damping may consist of tuned mass, sloshing, pendulum, or viscous damping systems.

5.2.4   Creep, Shrinkage, and Elastic Shortening

Vertical elements within the floor plan of tall towers are subjected to shortening. This shortening begins during the construction process and can continue up to 10,000 days after construction is complete. Vertical displacements affect non-structural components, including the exterior wall, interior partitions, and vertical MEP (mechanical/electrical/plumbing) systems. These displacements

Construction Image,
Jin Mao Tower, Shanghai, China

Construction Image,
Jin Mao Tower, Shanghai, China

are known to cause failure in supports for water piping within tall structures where creep and shrinkage, particularly in concrete towers, occur long after the pipes have been placed. Vertical displacements of 300 mm (12 in) or more at the top of the structure are not uncommon. Relative displacements between vertical elements on a floor plate can cause sloped floors or overloading of interconnection elements not designed for the relative movement.

Steel structures are more easily predictable when evaluating long-term relative movement between vertical elements within the structural system. Their displacements are only affected by elastic shortening due to axial load, assuming that the vertical elements are not permanently affected by eccentric long-term loads. Therefore, the variables in considering relative displacement are the self-weight dead loads and their distribution to vertical elements, and superimposed dead and live loads after the structure is complete.

Relative displacements between vertical elements in reinforced concrete or composite structures are far more difficult to predict since time, geometry, material composition, curing, and load all contribute to short- and long-term creep, shrinkage, and elastic shortening. Equalization of applied stress to these elements is an important design consideration. For preliminary calculations, a total strain of 700×10^{-6} in/in can be used.

The following guidelines should be used in designing for creep, shrinkage, and elastic shortening:

1. Determine construction sequence.
2. Calculate design loads (only realistic, sustained loads must be considered) and attempt to consider the actual compressive strength anticipated for the structure (in most cases, the actual in-place concrete compressive strength could be 10–25% higher than the theoretical design value). Over 90% of the sustained load in central reinforced shear wall systems is likely to be dead plus superimposed dead load. Of the total load on an exterior steel column, 75% is likely to be dead load plus superimposed dead load, 15% exterior wall load, and 10% live load.
3. Calculate anticipated vertical displacements.
4. Develop a construction program that requires the contractor to build the structure to "design elevation." This can be done with laser surveying techniques and making modifications to structural elements during construction (i.e. adjusting formwork heights, providing prefabricated shims for structural steel).
5. Establish displacements at the time of construction of elements at a particular floor elevation. This method allows for "zeroing out" of displacements within the system to this point.
6. Determine corrections that are required to control relative displacements between elements. It is important not to overcompensate, remembering that an element placed higher than the design elevation must "travel" through a theoretical zero or level point and can be located below a relative member and still be within acceptable tolerances.

Elastic Strain Equation at time i:

$$\varepsilon_{e_i} = \frac{P_{g_i}}{A_{t_i} E_{c_i}} + \varepsilon_{e_{i-1}}$$

where,

ε_{e_i} = total elastic strain at time i

P_{g_i} = incremental gravity load applied at time i (kN)

A_{t_i} = transformed section area at time i (mm^2)

E_{c_i} = concrete modulus of elasticity at time i (MPa)

$\varepsilon_{e_{i-1}}$ = total elastic strain at the previous time interval

Shrinkage Strain Equation at time i:

$$\varepsilon_{e_i} = (\varepsilon_{s_i w} - \varepsilon_{s_{i-1} w}) R_{cf_i} + \varepsilon_{e_{i-1}}$$

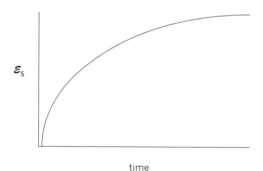

where,

ε_{s_i} = total shrinkage strain at time i

$\varepsilon_{s_i w}$ = baseline shrinkage strain at time i

$\varepsilon_{s_{i-1} w}$ = baseline shrinkage strain at the previous time interval

R_{cf_i} = steel reinforcement correction factor at time i

$\varepsilon_{s_{i-1}}$ = total shrinkage strain at the previous time interval

Incremental Creep Strain Equation at time *i* due to load at time *j*:

$$\varepsilon_{c_{(i-1)\rightarrow ij}} = (\varepsilon_{c_i w_j} - \varepsilon_{c_{(i-1)} w_j}) R_{cf_i}$$

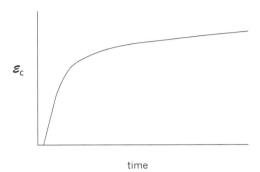

where,

$\varepsilon_{c_{(i-1)\rightarrow ij}}$ = incremental creep strain at time *i* due to load at time *j*

$\varepsilon_{c_i w_j}$ = creep strain at time *i* due to load at time *j* without rebar effect

$\varepsilon_{c_{(i-1)} w_j}$ = creep strain at the previous time interval due to load at time *j* without rebar effect

R_{cf_i} = steel reinforcement correction factor at time *i*

CHAPTER 6

CHARACTERISTICS

DYNAMIC PROPERTIES, aerodynamics, placement of structural materials, floor-to-floor heights, and aspect ratios are all important characteristics in maximizing the structural efficiency of the tall tower.

6.1 DYNAMIC PROPERTIES

The fundamental period of the tall tower roughly can be estimated by considering the number of anticipated stories divided by 10. According to the ANSI (ASCE-88) Code, the following periods may be calculated for translational and torsional responses of tall structures:

Steel buildings: $T = 0.085H^{0.75}$

Concrete buildings: $T = 0.061H^{0.75}$

Steel or concrete buildings: $T_\theta = 0.054N$

where,

T = fundamental translational period

T_θ = first torsional period

H = building height in meters

N = number of stories

FACING PAGE

Al Hamra Tower Under Construction, Kuwait City, Kuwait

Building (HT, Material)	Fundamental Trans. Period (sec.)	Torsional Period (sec.)
Jin Mao Tower (H = 421 m, mixed)	$T_1 = 5.7$, $T_2 = 5.7$	$T_\theta = 2.5$
Burj Khalifa (H = 828 m, r/c)	$T_1 = 11.0$, $T_2 = 10.0$	$T_\theta = 4.0$
Al Hamra Tower (H = 415 m, r/c)	$T_1 = 7.5$, $T_2 = 5.9$	$T_\theta = 3.2$
Goldfield Int'l Garden (H = 150 m, r/c)	$T_1 = 4.4$, $T_2 = 3.6$	$T_\theta = 2.4$
Jinta Tower (H = 369 m, steel)	$T_1 = 8.2$, $T_2 = 7.5$	$T_\theta = 6.1$
Jinao Tower (H = 235 m, mixed)	$T_1 = 5.0$, $T_2 = 4.8$	$T_\theta = 3.6$

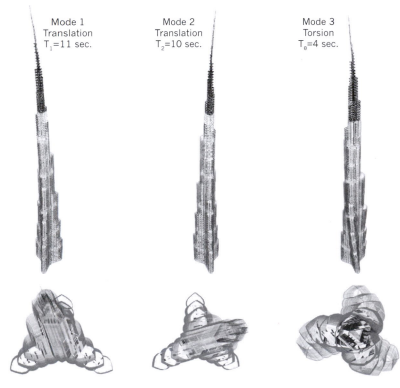

Mode 1
Translation
$T_1 = 11$ sec.

Mode 2
Translation
$T_2 = 10$ sec.

Mode 3
Torsion
$T_\theta = 4$ sec.

Dynamic Behavior,
Burj Khalifa, Dubai, UAE

6.2 AERODYNAMICS

The aerodynamics of the tall tower are important in minimizing imposed design forces. Across-wind motion (perpendicular to the direction of applied wind load) governs the behavior of the structure. Organized vortex shedding generates the highest force levels. Vortex shedding is most organized (has the most adverse effects) with circular cross-sectional shapes, less organized with triangular shapes, and least organized with square shapes. The introduction of holes through the building cross-sections improves the behavior further. Variation of the structure's cross-section along its height also acts to disorganize or disseminate vortex shedding. Strouhal numbers describe oscillating flow mechanisms and are based on the frequency of vortex shedding, shape geometry, and wind velocity. Strouhal numbers for common building shapes are shown below in parenthesis.

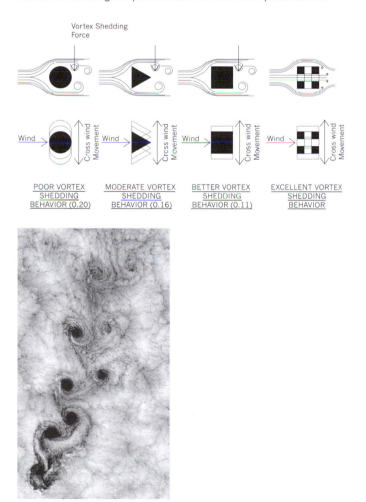

Vortex Shedding

6.3 PLACEMENT OF STRUCTURAL MATERIALS

Placement of structural materials within the tall tower is crucial to efficiency and economy. Placing material at the perimeter of the structure leads to the highest effective stiffness. Large material concentrations (columns) at the four corners of a square plan are most efficient, with only 50% efficiency realized with concentrated materials at the midpoint of the face of a square or distributed evenly along a circular plan.

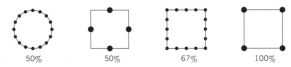

50%   50%   67%   100%

6.4 FLOOR-TO-FLOOR HEIGHTS

Minimizing floor-to-floor heights maximizes use within the structure. Every effort should be made to coordinate building systems and clear ceiling heights. The desired number of floors can be achieved while minimizing the building height, or the number of floors used within a specified height limit can be maximized. For office use, a 2.75 m (9'-0") tall ceiling height nominally yields a floor-to-floor height of 4.0 m (13'-1½"), and for residential use

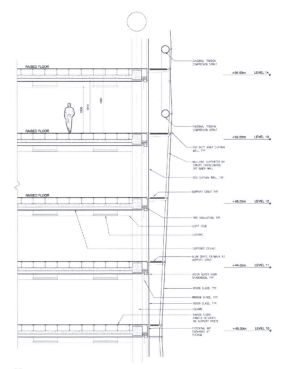

Typical Floor-to-Floor Tower Section, Jinao Tower, Nanjing, China

a nominal floor-to-floor height of 3.2 m (10'-6") can be used for the basis of design with a minimum finished ceiling height of 2.4 m (8'-0") (could be considerably higher without drop ceiling).

6.5   ASPECT RATIOS

The aspect ratio of a building is the ratio of height to the structural width of the base of the building. For super-tall buildings, it is essential to activate the full structural width of the building. With structural systems located at the perimeter of the building, a target aspect ratio of between 6 and 7 to 1 is common. For tall towers, this ratio may be 8 to 1 or more. Typically, buildings with aspect ratios of greater than 8 to 1 should consider supplemental damping systems to reduce occupant perception of motion. Shear wall cores centrally located typically have aspect ratios of between 10 and 15 to 1.

Building	Height	Aspect Ratio (height/width)	Material
Willis Tower (formerly Sears Tower)	445 m	6.4	Steel
Jin Mao Tower	421 m	7.0	Mixed
Amoco Building	346 m	6.0	Steel
John Hancock	344 m	6.6	Steel

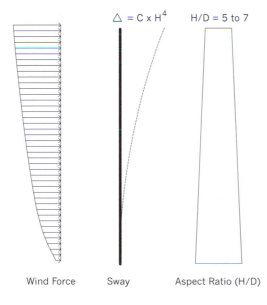

$$\triangle = C \times H^4 \qquad H/D = 5 \text{ to } 7$$

Wind Force   Sway   Aspect Ratio (H/D)

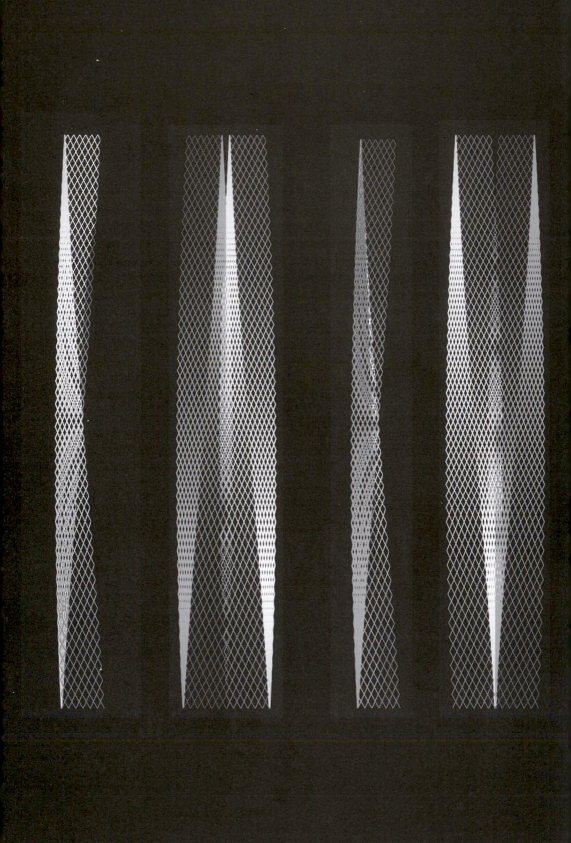

CHAPTER 7
SYSTEMS

THERE ARE SEVERAL FACTORS to consider when selecting a structural system for tall buildings. Safety, occupant comfort, and economy are the most important. The use and aesthetic of the structure also dictate possible solutions. For instance, it may be desirable to have wide column spacing at the perimeter rather than closely spaced columns. Available construction materials, available construction time, and contractor expertise must also be considered. Site factors such as poor soil conditions or high seismicity may prescribe a particular structural system.

Building use is an important consideration when selecting a structural system. For instance, residential (or hotel) construction generally aligns well with a reinforced concrete structure. This structural system allows for smaller spans and in many cases flat plate (slab) floor systems. Post-tensioning can be used to increase spans while maintaining a minimal structural framing depth. Generally, a 9 m × 9 m (30 ft × 30 ft) column spacing works well for residential use and can be readily coordinated with parking requirements that typically exist below the residential structure. The underside of this framing system can be finished and painted, serving as an acceptable finished ceiling while maximizing usable height for the spaces. A structural steel or composite system is also appropriate for this type of construction, and may be best if construction speed is an issue, but requires fire-proofing and finished ceilings.

Structural System Sketches,
Jinling Tower Competition, Nanjing, China

FACING PAGE
Jinling Tower Competition Rendering,
Nanjing, China

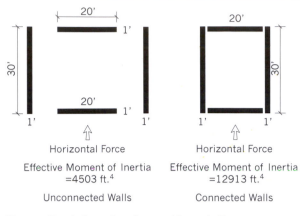

Horizontal Force

Effective Moment of Inertia
=4503 ft.^{4}

Unconnected Walls

Horizontal Force

Effective Moment of Inertia
=12913 ft.^{4}

Connected Walls

Moment of Inertia Comparison, Structural System in Plan

Reinforced concrete systems can also be used for office buildings, but long-span conditions should be carefully considered. Ideal office space configurations result in spans of approximately 13.5 m (45 ft), resulting in rather deep conventional reinforced concrete beam framing. Post-tensioning can be used in the beam framing to reduce overall depth. It is generally not advised to use post-tensioning in floor slabs for an office building since tenant modifications (interconnecting stairs etc.) over the life of the structure

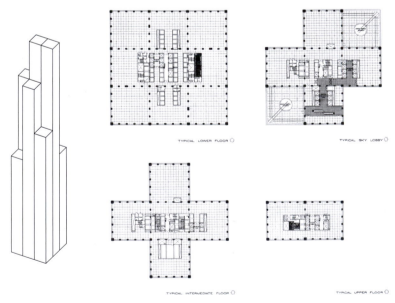

Tubular Structural System Concept and Resulting Floor Plans,
Willis Tower (formerly Sears Tower), Chicago, IL

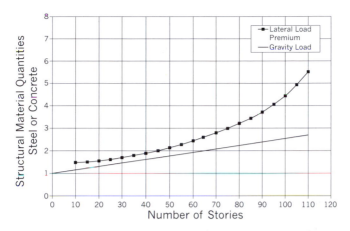

Structural Material Quantity vs. Number of Stories

are difficult to incorporate as post-tensioning tendons cannot be cut. Designated areas where conventional reinforcing could be used and raised floors (services are distributed through raised floor rather than through structure) generally alleviate these limitations.

Structural steel or composite systems can be used for either residential or office construction. For residential uses, the floor-to-floor heights generally increase (finished ceilings required underneath steel framing systems) and typically ceilings are coffered around framing to give maximum ceiling heights. For office buildings, steel framing works well, allowing for future modifications (local areas generally can be re-framed to accommodate openings etc.) and long-span conditions can be readily addressed. Wide-flanged framing with mechanical systems distributed below the beams, or built-up truss framing with mechanical systems distributed through the trusses, can be used. Framing is fire-proofed with finished ceilings generally required. Raised floor systems can be incorporated to allow for mechanical air distributions, as well as electrical and data distributions.

7.1   MATERIAL QUANTITIES

In general, structural materials required to resist gravity loads are fairly constant for low-to-mid-rise structures, but increase linearly for taller structures. In addition, the amount of material required for the lateral load-resisting system increases significantly with height.

A building is usually considered a "high-rise" if its height exceeds 23 m (75 ft). At extreme heights of 610 m (2000 ft) and above, increasingly complex design considerations must be taken into account.

Generally, the minimum amount of material required for gravity loads is as follows (apply quantities to gross framed building area):

Steel building: 49 kg/m² (10 psf)
                                  0.1 m³/m² (0.4 cu ft/sf) (concrete)*
                                  4.8 kg/m² (1.0 psf) (rebar)*

Reinforced concrete building: 32 kg/m² (6.5 psf) (rebar)
                                  0.34 m³/m² (1.1 cu ft/sf) (concrete)

*Concrete and rebar required typically for metal deck slab systems in steel buildings.

As heights increase, the considerations for structural systems change. For instance, the steel frame used in the Reliance Building would require approximately 68 kg/m² (14 psf) of structural steel for the 15-story building height of 61.6 m (202 ft). If this same building frame were used in a building of 50 stories (approx. 200 m or 656 ft), the material required for strength and stiffness of the moment-resisting frame may be 244 kg/m² (50 psf) or more. On the other hand, if a braced frame were used in the core area and a steel frame at the perimeter, the material quantities required may be reduced to 122 kg/m² (25 psf).

7.2   PRACTICAL LIMITS OF STRUCTURAL SYSTEMS

As the number of stories and structural heights increase, structural solutions must respond to behavior typically controlled by lateral wind loads (and gravity). In regions of high seismicity, ductile detailing of the system components is crucial to successful performance. In many cases, lateral loads imposed by seismic ground motions can govern the design (strength or serviceability) based on mass and geometric and stiffness characteristics.

The John Hancock Center . . . was designed in six months' less time by computers.

X-Braces Trim Steel Tonnage

The Genius of Dr. Fazlur Khan—Innovative Structural Systems and the Use of Main Frame Computing

When structural systems are defined for a tower structure, it is usually implied that the system describes the lateral load-resisting portion of the structure. The lateral system typically serves dual roles, resisting wind and seismic forces as well as resisting gravity loads. Many of these structures have additional structural elements that resist gravity loads only, including floor slabs, floor framing beams, and columns that support framing with shear or pin connections only.

Limits of structural systems based on the number of stories, and therefore height, are subjective, but based on years of development have proven to lead to the most efficient and safe solutions. For instance, a

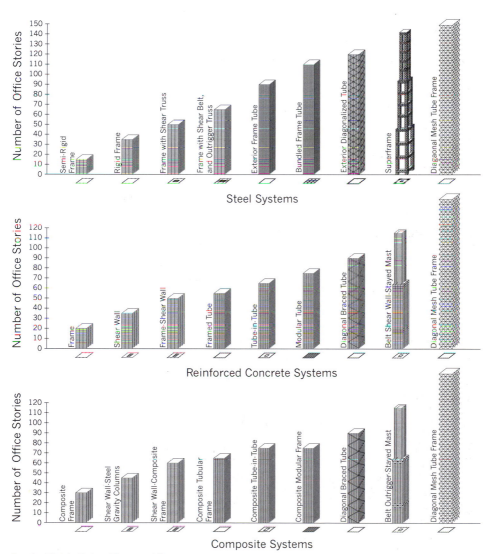

Practical Height Limits of Structural Systems

moment-resisting rigid frame may have a practical height limit of 35 stories or 140 m (460 ft), but can and has been used in structures of greater height (the Empire State Building is a good example of where a rigid moment frame was used for a 102 story, 382 m (1252 foot) tall structure). In each case the proposed structural solution must be superimposed and coordinated with the proposed architecture.

Practical limits of structural systems are shown in the following diagrams. The systems are typically associated with office building floor-to-floor heights, but can similarly be applied to other building uses.

7.2.1   Structural Steel

The building heights listed below are based on an average floor-to-floor height of 4.0 m (13'-1½"), allowing for a 2.75 m (9'-0") tall ceiling. Six meters (20 ft) are included for lobby spaces. One 8 meter (26 ft) tall space is included in the mid-height of the tower for mechanical spaces or a sky lobby for buildings 60 stories and above.

7.2.1.1   Steel Semi-Rigid Frame

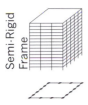

General limit of number of stories = 15 and height = 62 meters (203'-4")

The moment-resisting frame is typically comprised of wide-flanged shaped columns and beams with primary frame connections typically bolted with slip-critical bolts. Frame connections allow for partial fixity, permitting some rotation when loaded. Column spacing typically ranges from approximately the floor-to-floor height to twice the floor-to-floor height (generally, spacing varies between 4.5 m (15 ft) and 9 m (30 ft)). The taller the building, the greater the need to use deeper wide-flanged sections for beams and columns and closer column spacing. Columns typically vary from W14 to W36 sections and beams typically vary from W21 to W36 sections.

7.2.1.2   Steel Rigid Frame

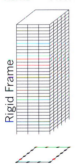

General limit of number of stories = 35 and height = 142 meters (465'-9")

The moment-resisting frame is typically comprised of wide-flanged shaped columns and beams with primary frame connections typically bolted with slip-critical bolts, welded, or a combination of welded/bolted. Frame connections are fully fixed, developing the full bending moment capacity of the beams. Column spacing typically ranges from approximately the floor-to-floor height to twice the floor-to-floor height; generally, spacing varies between 4.5 m (15 ft) and 9 m (30 ft). The taller the building, the greater the need to use deeper wide-flanged sections for beams and columns and closer column spacing. Columns typically vary from W14 to W36 sections and beams typically vary from W21 to W36 sections.

7.2.1.3　Steel Frame with Shear Truss

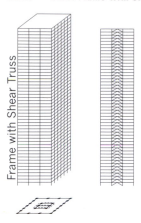

General limit of number of stories = 50 and height = 202 meters (662'-6")

The shear truss, typically consisting of diagonal wide-flanged, angle, T-shaped, or tubular members, is located in the central core area or at the perimeter. Chevron (or k-braced) or concentrically braced (x-braced) frames are typically used. Gusset plates are typically used to connect intersecting members. Columns are typically spaced 4.5 m (15 ft) to 9 m (30 ft) on-center.

The frame is typically comprised of wide-flanged shaped columns and beams with primary frame connections typically bolted, welded, or a combination of welded/bolted. Frame connections are fully fixed, developing the full bending moment capacity of the beams. Column spacing typically ranges from approximately the floor-to-floor height to twice the floor-to-floor height; generally, spacing varies between 4.5 m (15 ft) and 9 m (30 ft). The taller the building, the greater the need to use deeper wide-flanged sections for beams and columns and closer column spacing. Columns typically vary from W14 to W36 sections and beams typically vary from W21 to W36 sections.

Shear truss is typically used to resist lateral loads for strength and drift control. Moment-resisting frames are used to supplement lateral load strength and drift control, providing increased torsional resistance of the structural system.

7.2.1.4   Steel Frame with Shear, Belt, and Outrigger Trusses

Frame with Shear, Belt, and Outrigger Truss

General limit of number of stories = 65 and height = 266 meters (872'-6")

The shear truss, typically consisting of diagonal wide-flanged, angle, T-shaped, or tubular members, is located in the central core area or at the perimeter. Chevron (or k-braced) or concentrically braced (x-braced) frames are typically used. Gusset plates are typically used to connect intersecting members. Columns are typically spaced 4.5 m (15 ft) to 9 m (30 ft) on-center.

Outrigger trusses, typically at least one story tall (preferably two stories tall), are used to interconnect the central core and perimeter frame. Outrigger trusses typically consist of large wide-flanged or built-up members.

Belt trusses, located within the perimeter frame at the same level as the outrigger trusses, are used to distribute forces fairly evenly from the outrigger trusses to the perimeter frame.

The frame is typically comprised of wide-flanged shaped columns and beams with primary frame connections typically bolted, welded, or a combination of welded/bolted. Frame connections are fully fixed, developing the full bending moment capacity of the beams. Column spacing typically ranges from approximately the floor-to-floor height to twice the floor-to-floor height (generally, spacing varies between 4.5 m (15 ft) and 9 m (30 ft)). The taller the building, the greater the need to use deeper wide-flanged sections for beams and columns and closer column spacing. Columns typically vary from W14 to W36 sections and beams typically vary from W21 to W36 sections.

Shear truss is typically used to resist lateral loads for strength and drift control. Moment-resisting frames are used to supplement lateral load strength and drift control, providing increased torsional resistance of the structural system. Outrigger trusses, generally located at 25% of the height, 50% of the height, and top of the structure, act as levers, restricting the displacements of the core and introducing axial loads into columns of the perimeter frame.

7.2.1.5   Steel Exterior Framed Tube

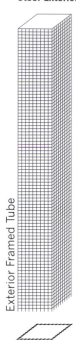

Exterior Framed Tube

General limit of number of stories = 90 and height = 366 meters (1200'-6")

The tubular frame is typically comprised of wide-flanged shaped or built-up columns and beams with primary frame connections typically bolted, welded, or a combination of welded/bolted. Frame connections are fully fixed, developing the full bending moment capacity of the beams. Column spacing approximately matches the floor-to-floor height, while aligning with interior partitions/exterior wall mullion modules (column spacing typically 4.5 m (15 ft)). The taller the building, the greater the need to use deeper wide-flanged sections for beams and columns and potentially closer column spacing.

The tubular frame is engineered to approximately equalize the bending stiffness of the columns and the beams. The structural system attempts to equalize axial load (reduce shear lag) attracted to columns when the overall structure is subjected to lateral loads. Shear lag is the phenomenon that describes the unequal distribution of force on the leading or back face columns when the frame is subjected to lateral loads.

7.2.1.6   Steel Bundled Frame Tube

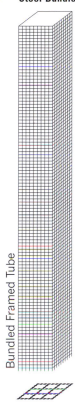

Bundled Framed Tube

General limit of number of stories = 110 and height = 446 meters (1462'-10")

The tubular frame is typically comprised of wide-flanged shaped or built-up columns and beams with primary frame connections typically bolted, welded, or a combination of welded/bolted. Frame connections are fully fixed, developing the full bending moment capacity of the beams. Column spacing approximately matches the floor-to-floor height, while aligning with interior partitions/exterior wall mullion modules (column spacing typically 4.5 m (15 ft)). The taller the building, the greater the need to use deeper wide-flanged sections for beams and columns and potentially closer column spacing.

The tubular frame is engineered to approximately equalize the bending stiffness of the columns and the beams. The structural system attempts to equalize axial load (reduce shear lag) attracted to columns when the overall structure is subjected to lateral loads. Shear lag is the phenomenon that describes the unequal distribution of force on the leading or back face columns when the frame is subjected to lateral loads.

The bundled tube uses a cellular concept, introducing interior frames, to further reduce shear lag. These bundle tubes may include belt trusses at levels where floor plans transition from large to small in order to interconnect or tie the tubular frames together.

7.2.1.7   Steel Exterior Diagonal Tube

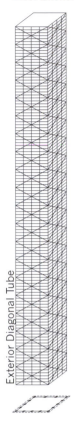

Exterior Diagonal Tube

General limit of number of stories = 120 and height = 486 meters (1594'-1")

The exterior diagonal tube is typically comprised of wide-flanged shaped or built-up columns and beams with primary frame connections typically bolted, welded, or a combination of welded/bolted. Frame connections are fully fixed, developing the full bending moment capacity of the beams. Column spacing approximately matches the floor-to-floor height, while aligning with interior partitions/exterior wall mullion modules (column spacing typically 4.5 m (15 ft)). The taller the building, the greater the need to use deeper wide-flanged sections for beams and columns and potentially closer column spacing.

The exterior diagonal tube introduces diagonal members into the perimeter of the structure. These diagonal members typically exist on multiple-floor intervals. The structural system attempts to equalize axial load (reduce shear lag) attracted to columns when the overall structure is subjected to lateral loads. Shear lag is the phenomenon that describes the unequal distribution of force on the leading or back face columns when the frame is subjected to lateral loads. The use of diagonal members in the tube significantly increases structural efficiency (decreases the use of materials) since the behavior is dominated by axial rather than bending behavior.

7.2.1.8   Steel Superframe

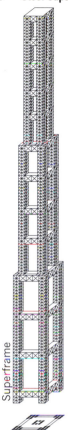

Superframe

General limit of number of stories = 130 and height = 526 meters (1725'-3")

The exterior diagonal frame is typically comprised of wide-flanged shaped or built-up columns and beams with primary frame connections typically bolted, welded, or a combination of welded/bolted. Frame connections are fully fixed, developing the full bending moment capacity of the beams. Column spacing approximately matches the floor-to-floor height, while aligning with interior partitions/exterior wall mullion modules (column spacing typically 4.5 m (15 ft)). The taller the building, the greater the need to use deeper wide-flanged sections for beams and columns and potentially closer column spacing.

The perimeter frame is configured on a mega or superframe module with horizontal and vertical members of the frame existing in multiple bays and multiple stories. Each bay of the perimeter frame is diagonalized. The system allows for plan transitions along the height, and incorporates inherent belt trusses at transition floors. The concept allows for large atria or openings through the structure and also allows for winds to pass through the structure to reduce the shear and overturning demand on the structure.

7.2.1.9   Steel Diagonal Mesh Tube Frame

Diagaonal Mesh Tube Frame

General limit of number of stories = 150 and height = 606 meters (1987'-7")

The exterior diagonal mesh tube frame is typically comprised of wide-flanged shaped or built-up columns and beams with primary frame connections typically bolted, welded, or a combination of welded/bolted. Frame connections are fully fixed, developing the full bending moment capacity of the beams. Column spacing approximately matches the floor-to-floor height while aligning with interior partitions/exterior wall mullion modules (column spacing typically 4.5 m (15 ft)). The taller the building, the greater the need to use deeper wide-flanged sections for beams and columns and potentially closer column spacing.

The perimeter frame is configured to introduce only axial load into diagonal frame members. Bending moments on individual members are essentially eliminated, significantly increasing structural efficiency while minimizing material quantities. Tighter weave of the mesh results in smaller structural members and even greater efficiency.

7.2.2 Reinforced Concrete

The building heights listed below are based on an average floor-to-floor height of 3.2 m (10'-6"), allowing for a 2.45 m (8'-0") tall ceiling. Six meters (20 ft) are included for lobby spaces. One 6.4 meter (21 ft) tall space is included in the mid-height of the tower for mechanical spaces or a sky lobby for buildings 60 stories and above.

7.2.2.1 Concrete Frame

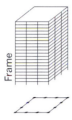

General limit of number of stories = 20 and height = 66.8 meters (219'-1")

The moment-resisting frames are typically comprised of rectangular or square columns and rectangular beams. Frame joints use rebar detailing to develop the full bending moment capacity of the beams. Column spacing typically ranges from approximately the floor-to-floor height to twice the floor-to-floor height (generally, spacing varies between 4.5 m (15 ft) and 9 m (30 ft)). The taller the building, the greater the need to use deeper rectangular sections for beams and columns and closer column spacing, with rectangular column sections orientated to provide greatest bending resistance.

7.2.2.2   Concrete Shear Wall

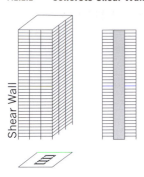

Shear Wall

General limit of number of stories = 35 and height = 114.8 meters (376'-6")

Shear walls within a tower are most commonly centrally located around service areas, including elevators, mechanical spaces, and rest-rooms. However, these walls may be distributed throughout the floor plan in structures with residential programs. In some cases shear walls are located eccentrically in the floor plan and must resist significant torsion due to eccentrically applied wind loads or seismic loads arising from the eccentric relationship between center of mass and lateral stiffness.

Shear wall spacing locations vary, but they are generally located 9 m (30 ft) apart to allow for a double bank of elevators and an elevator lobby. Link beams are used to interconnect wall segments where doors or mechanical openings are required in the core. The link beam depths are generally maximized to obtain greatest shear and bending resistance and must be coordinated with doorway heights and mechanical systems.

Shear walls resist all lateral loads and can be subjected to net tension and foundation uplift/overturning. Therefore, plan location/size and gravity load balancing is important to minimize any tension on shear wall elements.

7.2.2.3   Concrete Frame – Shear Wall

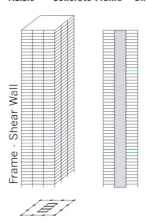

General limit of number of stories = 50 and height = 162.8 meters (534'-0")

Shear walls within a tower are most commonly centrally located around service areas, including elevators, mechanical spaces, and restrooms. However, these walls may be distributed throughout the floor plan in structures with residential programs. In some cases, shear walls are located eccentrically in the floor plan and must resist significant torsion due to eccentrically applied wind loads or seismic loads arising from the eccentric relationship between center of mass and lateral stiffness.

Shear wall spacing locations vary, but they are generally located 9 m (30 ft) apart to allow for a double bank of elevators and an elevator lobby. Link beams are used to interconnect wall segments where doors or mechanical openings are required in the core. The link beam depths are generally maximized to obtain greatest shear and bending resistance and must be coordinated with doorway heights and mechanical systems.

Frames are combined with the shear walls to increase strength and stiffness. The moment-resisting frames are typically comprised of rectangular or square columns and rectangular beams. Frame joints use rebar detailing to develop the full bending moment capacity of the beams. Column spacing typically ranges from approximately the floor-to-floor height to twice the floor-to-floor height (generally, spacing varies between 4.5 m (15 ft) and 9 m (30 ft)). The taller the building, the greater the need to use deeper rectangular sections for beams and columns and closer column spacing, with rectangular column sections orientated to provide greatest bending resistance.

Shear walls resist a majority of the lateral load, especially in lower portions of the structure, and can be subjected to net tension and foundation uplift/overturning. Therefore, plan location/size and gravity load balancing is important to minimize any tension on shear wall elements.

7.2.2.4 Concrete Framed Tube

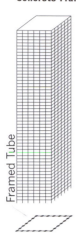

Framed Tube

General limit of number of stories = 55 and height = 178.8 meters (586'-6")

Moment-resisting tubular frames are typically comprised of rectangular or square columns and rectangular beams. Frame joints use rebar detailing to develop the full bending moment capacity of the beams. Column spacing typically ranges from a dimension slightly less than the floor-to-floor height to approximately equal to the floor-to-floor height (generally, spacing varies between 3.0 m (10 ft) and 4.5 m (15 ft)). The taller the building, the greater the need to use deeper rectangular sections for beams and columns and closer column spacing, with rectangular column sections orientated to provide greatest bending resistance.

The tubular frames are engineered to approximately equalize the bending stiffness of the columns and the beams. The structural system attempts to equalize axial load (reduce shear lag) attracted to columns when the overall structure is subjected to lateral loads. Shear lag is the phenomenon that describes the unequal distribution of force on the leading or back face columns when the frame is subjected to lateral loads.

7.2.2.5   Concrete Tube-in-Tube

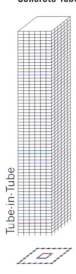

Tube-in-Tube

General limit of number of stories = 65 and height = 214 meters (701'-11")

Moment-resisting tubular frames are typically comprised of rect-angular or square columns and rectangular beams. Frame joints use rebar detailing to develop the full bending moment capacity of the beams. Column spacing typically ranges from a dimension slightly less than the floor-to-floor height to approximately equal to the floor-to-floor height (generally, spacing varies between 3.0 m (10 ft) and 4.5 m (15 ft)). The taller the building, the greater the need to use deeper rectangular sections for beams and columns and closer column spacing, with rectangular column sections orientated to provide greatest bending resistance.

The tubular frames are engineered to approximately equalize the bending stiffness of the columns and the beams. The structural system attempts to equalize axial load (reduce shear lag) attracted to columns when the overall structure is subjected to lateral loads. Shear lag is the phenom-enon that describes the unequal distribution of force on the leading or back face columns when the frame is subjected to lateral loads.

Interior tubular frames or shear walls are combined with perimeter tubular frames. These frames or shear walls provide additional strength and stiffness to the tubular system.

7.2.2.6   Concrete Modular Tube

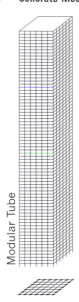

Modular Tube

General limit of number of stories = 75 and height = 246 meters (806'-10")

Moment-resisting tubular frames are typically comprised of rect-angular or square columns and rectangular beams. Frame joints use rebar detailing to develop the full bending moment capacity of the beams. Column spacing typically ranges from a dimension slightly less than the floor-to-floor height to approximately equal to the floor-to-floor height (generally, spacing varies between 3.0 m (10 ft) and 4.5 m (15 ft)). The taller the building, the greater the need to use deeper rectangular sections for beams and columns and closer column spacing, with rectangular column sections orientated to provide greatest bending resistance.

The tubular frames are engineered to approximately equalize the bending stiffness of the columns and the beams. The structural system attempts to equalize axial load (reduce shear lag) attracted to columns when the overall structure is subjected to lateral loads. Shear lag is the phenom-enon that describes the unequal distribution of force on the leading or back face columns when the frame is subjected to lateral loads.

Interior tubular frames are combined with perimeter tubular frames. These frames are configured to develop a cellular configuration in plan. Inte-rior frames must be carefully coordinated with occupied spaces since column spacing is limited on frame lines. These frames provide additional strength and stiffness to the tubular system.

7.2.2.7   Concrete Diagonal Braced Tube

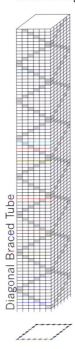

Diagonal Braced Tube

General limit of number of stories = 90 and height = 294 meters (964'-4")

Moment-resisting frames are typically comprised of rectangular or square columns and rectangular beams. Frame joints use rebar detailing to develop the full bending moment capacity of the beams. Column spacing typically ranges from approximately the floor-to-floor height to twice the floor-to-floor height (generally, spacing varies between 3.0 m (10 ft) and 4.5 m (15 ft)). The taller the building, the greater the need to use deeper rectangular sections for beams and columns and closer column spacing, with rectangular column sections orientated to provide greatest bending resistance.

The tubular frames are infilled with a diagonal system at exterior locations. The structural system attempts to equalize axial load (reduce shear lag) attracted to columns when the overall structure is subjected to lateral loads. Shear lag is the phenomenon that describes the unequal distribution of force on the leading or back face columns when the frame is subjected to lateral loads.

7.2.2.8   Concrete Belt Shear Wall-Stayed Mast

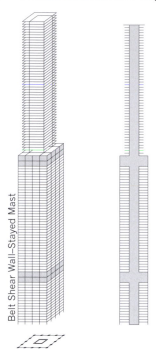

Belt Shear Wall–Stayed Mast

General limit of number of stories = 110 and height = 358 meters (1174'-3")

Mega-core shear walls are centrally located in the floor plan around service areas, including elevators, mechanical spaces, and restrooms. A closed form (square, rectangular, circular, octagon-shaped) is generally required to provide translational and torsional resistance.

Shear wall spacing locations vary, but they are generally located 9 m (30 ft) apart to allow for a double bank of elevators and an elevator lobby. Link beams are used to interconnect wall segments where doors or mechanical openings are required in the core. The link beam depths are generally maximized to obtain greatest shear and bending resistance and must be coordinated with doorway heights and mechanical systems.

Mega-core shear walls are interconnected to perimeter columns or frames with concrete walls (two stories tall typical) or in some cases steel trusses. These outrigger walls transfer loads from the central core to the perimeter columns or frames and maximize the structural depth of the floor plan.

The mega-core shear walls resist a majority of the lateral load, especially in lower portions of the structure, and can be subjected to net tension and foundation uplift/overturning. Therefore, plan location/size and gravity load balancing is important to minimize any tension on shear wall elements.

Belt walls may be used at outrigger levels to interconnect perimeter frame columns to maximize the axial stiffness of columns participating in the lateral system.

7.2.2.9   Concrete Diagonal Mesh Tube Frame

General limit of number of stories = 120 and height = 390 meters (1279'-2")

The exterior diagonal mesh tube frame is typically comprised of rectangular or square diagonal members. Rectangular reinforced concrete frame beams are generally required at floor levels. Diagonal member spacing approximately matches the floor-to-floor height, while aligning with interior partitions/exterior wall mullion modules (diagonal spacing typically 4.5 m (15 ft)). The taller the building, the greater the need to use deeper diagonal sections and frame beams. A closer diagonal spacing is optional (3.0 m (10 ft)).

The perimeter frame is configured to introduce only axial load into diagonal frame members. Bending moments on individual members are essentially eliminated, significantly increasing structural efficiency while minimizing material quantities. Tighter weave of the mesh results in smaller structural members and even greater efficiency.

7.2.3   Composite (Combination of Steel and Concrete)

The building heights listed below are based on an average floor-to-floor height of 4.0 m (13'-1½"), allowing for a 2.75 m (9'-0") tall ceiling. Six meters (20 ft) are included for lobby spaces. One 8 meter (26 ft) tall space is included in the mid-height of the tower for mechanical spaces or a sky lobby for buildings 60 stories and above.

7.2.3.1   Composite Frame

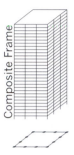

General limit of number of stories = 30 and height = 122 meters (400'-2")

   The moment-resisting frame is typically comprised of wide-flanged shaped columns and beams encased in reinforced concrete with primary frame connections typically bolted, welded, or a combination of welded/ bolted and rebar detailed to develop full fixity and bending moment capacity of the beams. Columns and beams are typically encased in rectangular or square concrete sections. Column spacing typically ranges from approximately the floor-to-floor height to twice the floor-to-floor height (generally, spacing varies between 4.5 m (15 ft) and 9 m (30 ft)). The taller the building, the greater the need to use deeper wide-flanged sections for beams and columns and closer column spacing. Columns typically vary from W14 to W36 sections and beams typically vary from W21 to W36 sections. Structural steel within concrete sections varies from 3% to 10% of the concrete section.

7.2.3.2   Concrete Shear Wall – Steel Gravity Columns

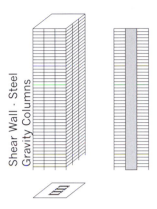

General limit of number of stories = 45 and height = 182 meters (597'-0")

Shear walls within a tower are most commonly centrally located around service areas, including elevators, mechanical spaces, and restrooms. However, these walls may be distributed throughout the floor plan in structures with residential programs. In some cases shear walls are located eccentrically in the floor plan and must resist significant torsion due to eccentrically applied wind loads or seismic loads arising from the eccentric relationship between center of mass and lateral stiffness.

Shear walls resist all lateral loads and can be subjected to net tension and foundation uplift/overturning. Therefore, plan location/size and gravity load balancing is important to minimize any tension on shear wall elements. Steel columns resist gravity loads only from composite steel floor framing.

7.2.3.3   Concrete Shear Wall – Composite Frame

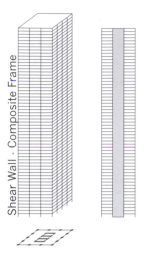

Shear Wall – Composite Frame

General limit of number of stories = 60 and height = 246 meters (806'-10")

Shear walls within a tower are most commonly centrally located around service areas, including elevators, mechanical spaces, and restrooms. However, these walls may be distributed throughout the floor plan in structures with residential programs. In some cases, shear walls are located eccentrically in the floor plan and must resist significant torsion due to eccentrically applied wind loads or seismic loads arising from the eccentric relationship between center of mass and lateral stiffness.

Shear wall spacing locations vary, but they are generally located 9 m (30 ft) apart to allow for a double bank of elevators and an elevator lobby. Link beams are used to interconnect wall segments where doors or mechanical openings are required in the core. The link beam depths are generally maximized to obtain greatest shear and bending resistance and must be coordinated with doorway heights and mechanical systems.

Frames are combined with the shear walls to increase strength and stiffness. The moment-resisting frame is typically comprised of wide-flanged shaped columns and beams encased in reinforced concrete, with primary frame connections typically bolted, welded, or a combination of welded/bolted and rebar detailed to develop full fixity and bending moment capacity of the beams. Columns and beams are typically encased in rectangular or square concrete sections. Frame joints use rebar detailing to develop the full bending moment capacity of the beams. Column spacing typically ranges from approximately the floor-to-floor height to twice the floor-to-floor height (generally, spacing varies between 4.5 m (15 ft) and 9 m (30 ft)). The taller the building, the greater the need to use deeper rectangular sections for beams and columns and closer column spacing, with rectangular column sections orientated to provide greatest bending resistance.

Shear walls resist a majority of the lateral load, especially in lower portions of the structure, and can be subjected to net tension and foundation uplift/overturning. Therefore, plan location/size and gravity load balancing is important to minimize any tension on shear wall elements.

7.2.3.4   Composite Tubular Frame

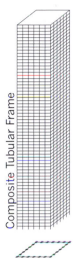

Composite Tubular Frame

General limit of number of stories = 65 and height = 266 meters (872'-6")

Moment-resisting tubular frames are typically comprised of structural steel wide-flange or built-up sections embedded in rectangular or square concrete columns and rectangular beams. Frame joints use bolted, welded, or a combination of bolted and welded steel connections and rebar detailing to develop the full bending moment capacity of the beams. Column spacing approximately matches the floor-to-floor height, while aligning with interior partitions/exterior wall mullion modules (column spacing typically 4.5 m (15 ft)). The taller the building, the greater the need to use deeper rectangular sections for beams and columns and closer column spacing, with rectangular column sections orientated to provide greatest bending resistance.

The tubular frames are engineered to approximately equalize the bending stiffness of the columns and the beams. The structural system attempts to equalize axial load (reduce shear lag) attracted to columns when the overall structure is subjected to lateral loads. Shear lag is the phenomenon that describes the unequal distribution of force on the leading or back face columns when the frame is subjected to lateral loads.

7.2.3.5   Composite Tube-in-Tube

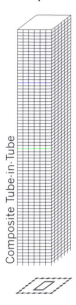

Composite Tube-in-Tube

General limit of number of stories = 75 and height = 306 meters (1003'-7")

Moment-resisting tubular frames are typically comprised of structural steel wide-flange or built-up sections embedded in rectangular or square concrete columns and rectangular beams. Frame joints use bolted, welded, or a combination of bolted and welded steel connections and rebar detailing to develop the full bending moment capacity of the beams. Column spacing approximately matches the floor-to-floor height, while aligning with interior partitions/exterior wall mullion modules (column spacing typically 4.5 m (15 ft)). The taller the building, the greater the need to use deeper rectangular sections for beams and columns and closer column spacing, with rectangular column sections orientated to provide greatest bending resistance.

The tubular frames are engineered to approximately equalize the bending stiffness of the columns and the beams. The structural system attempts to equalize axial load (reduce shear lag) attracted to columns when the overall structure is subjected to lateral loads. Shear lag is the phenomenon that describes the unequal distribution of force on the leading or back face columns when the frame is subjected to lateral loads.

Interior tubular frames or shear walls are combined with perimeter tubular frames. These frames or shear walls provide additional strength and stiffness to the tubular system.

7.2.3.6   Composite Modular Tube

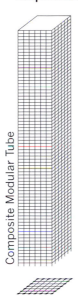

Composite Modular Tube

General limit of number of stories = 75 and height = 306 meters (1003'-7")

Moment-resisting tubular frames are typically comprised of structural steel wide-flange or built-up sections embedded in rectangular or square concrete columns and rectangular beams. Frame joints use bolted, welded, or a combination of bolted and welded steel connections and rebar detailing to develop the full bending moment capacity of the beams. Column spacing approximately matches the floor-to-floor height, while aligning with interior partitions/exterior wall mullion modules (column spacing typically 4.5 m (15 ft)). The taller the building, the greater the need to use deeper rectangular sections for beams and columns and closer column spacing, with rectangular column sections orientated to provide greatest bending resistance.

The tubular frames are engineered to approximately equalize the bending stiffness of the columns and the beams. The structural system attempts to equalize axial load (reduce shear lag) attracted to columns when the overall structure is subjected to lateral loads. Shear lag is the phenomenon that describes the unequal distribution of force on the leading or back face columns when the frame is subjected to lateral loads.

Interior tubular frames are combined with perimeter tubular frames. These frames are configured to develop a cellular configuration in plan. Interior frames must be carefully coordinated with occupied spaces since column spacing is limited on frame lines. These frames provide additional strength and stiffness to the tubular system.

117

7.2.3.7   Composite Diagonal Braced Tube

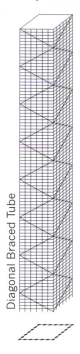

Diagonal Braced Tube

General limit of number of stories = 90 and height = 366 meters (1200'-6")

The exterior diagonal tube is typically comprised of structural steel wide-flange or built-up sections embedded in rectangular or square concrete columns and rectangular beams. Frame joints use bolted, welded, or a combination of bolted and welded steel connections and rebar detailing to develop the full bending moment capacity of the beams. Column spacing approximately matches the floor-to-floor height, while aligning with interior partitions/exterior wall mullion modules (column spacing typically 4.5 m (15 ft)). The taller the building, the greater the need to use deeper wide-flanged sections for beams and columns and potentially closer column spacing.

The exterior diagonal tube introduces diagonal members into the perimeter of the structure. These diagonal members typically exist on multiple-floor intervals. The structural system attempts to equalize axial load (reduce shear lag) attracted to columns when the overall structure is subjected to lateral loads. Shear lag is the phenomenon that describes the unequal distribution of force on the leading or back face columns when the frame is subjected to lateral loads. The use of diagonal members in the tube significantly increases structural efficiency (decreases the use of materials) since the behavior is dominated by axial rather than bending behavior.

7.2.3.8 Composite Belt Outrigger-Stayed Mast

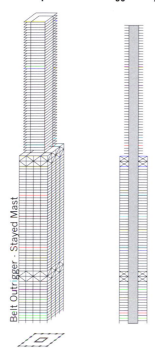

General limit of number of stories = 110 and height = 446 meters (1462'-10")

Mega-core shear walls are centrally located in the floor plan around service areas, including elevators, mechanical spaces, and restrooms. A closed form (square, rectangular, circular, octagon-shaped) is generally required to provide translational and torsional resistance.

Shear wall spacing locations vary, but they are generally located 9 m (30 ft) apart to allow for a double bank of elevators and an elevator lobby. Link beams are used to interconnect wall segments where doors or mechanical openings are required in the core. The link beam depths are generally maximized to obtain greatest shear and bending resistance and must be coordinated with doorway heights and mechanical systems.

Mega-core shear walls are interconnected to perimeter columns or frames with steel trusses (two stories tall typical). These outrigger trusses transfer loads from the central core to the perimeter composite columns or frames and maximize the structural depth of the floor plan.

The mega-core shear walls resist a majority of the lateral load, especially in lower portions of the structure, and can be subjected to net tension and foundation uplift/overturning. Therefore, plan location/size and gravity load balancing is important to minimize any tension on shear wall elements.

Belt trusses may be used at outrigger levels to interconnect perimeter frame columns to maximize the axial stiffness of columns participating in the lateral system.

7.2.3.9   Composite Diagonal Mesh Tube Frame

General limit of number of stories = 120 and height = 486 meters (1594'-1")

The exterior diagonal mesh tube is typically comprised of structural steel wide-flange or built-up sections embedded in rectangular or square diagonal concrete members. Rectangular composite frame beams are generally required at floor levels. Diagonal member spacing approximately matches the floor-to-floor height, while aligning with interior partitions/exterior wall mullion modules (diagonal spacing typically 4.5 m (15 ft)). The taller the building, the greater the need to use deeper diagonal sections and frame beams. A closer diagonal spacing is optional (3.0 m (10 ft)).

The perimeter frame is configured to introduce only axial load into frame members. Bending moments on individual members are essentially eliminated, significantly increasing structural efficiency while minimizing material quantities. Tighter weave of the mesh results in smaller structural members and even greater efficiency.

7.3   MAJOR SYSTEM DETAILS

Beyond typical details used for a majority of structural systems and components, unique system details not only solve key structural engineering problems but also potentially become defining elements of the architecture.

7.3.1   The Arch—Broadgate Phase II

The articulated structural steel joints at the base of the tied arch in Broadgate Phase II (Exchange House), London, located at each building face as well as two internal locations, resolve compression (arch) and tension (tie) forces. The connection results in a single, and essentially pure, reaction applying only vertical loads to reinforced concrete end piers and foundations below. The joints at the exterior faces are offset from the occupied spaces/exterior walls and are fire-engineered to prove that fire-proofing is not required and allow for exposed, painted structural steel.

7.3.2   The Rocker—The New Beijing Poly Plaza

A rocker or "reverse pulley" exists at Level 11 of the 22-story China Poly Headquarters Building in Beijing. The rocker allows for vertical hanging support of a museum through high-strength, spiral-wound bridge cables, while permitting the building to move freely during a seismic event. Without the rocker system, the large cables would act as tension braces, developing forces from large displacements of the top of the building relative to the top of the museum. If the cables acted as braces, the forces developed could not be accommodated by the cables or the base building structure. In addition, the rocker and cable system provide lateral support for the world's largest cable net (60 m × 90 m (197 ft × 295 ft)).

The Rocker,
New Beijing Poly Plaza, Beijing, China

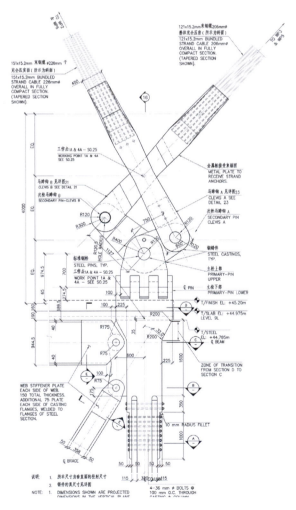

7.3.3   Pipe Collector—Jinao Tower

A reinforced concrete frame is "wrapped" with a conventional steel pipe system, allowing for a 40% reduction in the material that would otherwise be required for the lateral load-resisting system for the Jinao Tower, Nanjing, China. The pipes are eccentrically located at the corners of the structure, connected to steel elements encased in reinforced concrete columns. Since the diagonal pipe system only resists lateral load and the building is stable, even if part of the pipe system is lost to fire or another event, the pipes do not require fireproofing and can be simply painted and exposed (located in this structure within the double glass wall).

Pipe Collector,
Jinao Tower, Nanjing, China

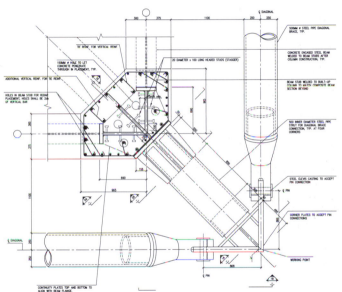

7.3.4   Pinned Trusses—Jin Mao Tower

Pins installed within massive steel outrigger trusses in the Jin Mao Tower, Shanghai, allowed for free movement during construction when creep, shrinkage and elastic shortening imposed differential settlement between the central reinforced concrete core and perimeter composite mega-columns. Initially conceived from a simple model of popsicle sticks, tongue depressors, and wood dowels, the trusses act as free moving mechanisms through the time of relative displacements until high-strength bolts are installed and the structure is placed in full service, resisting all lateral loads.

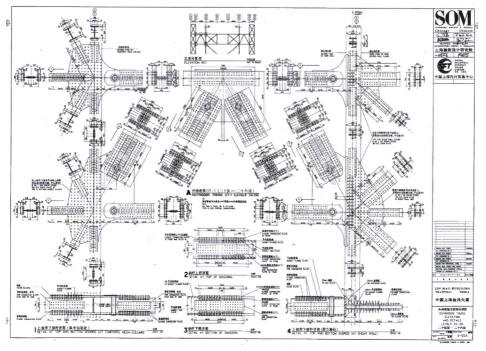

Working Drawing for Outrigger Truss,
Jin Mao Tower, Shanghai, China

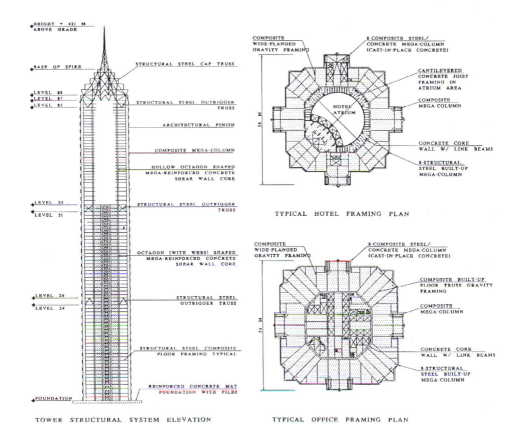

TOWER STRUCTURAL SYSTEM ELEVATION

Labels (elevation, left to right):
- HEIGHT = 421 M ABOVE GRADE
- BASE OF SPIRE
- STRUCTURAL STEEL CAP TRUSS
- LEVEL 88
- LEVEL 87
- LEVEL 85
- STRUCTURAL STEEL OUTRIGGER TRUSS
- ARCHITECTURAL FINISH
- COMPOSITE MEGA-COLUMN
- HOLLOW OCTAGON SHAPED MEGA-REINFORCED CONCRETE SHEAR WALL CORE
- LEVEL 53
- LEVEL 51
- STRUCTURAL STEEL OUTRIGGER TRUSS
- OCTAGON (WITH WEBS) SHAPED MEGA-REINFORCED CONCRETE SHEAR WALL CORE
- LEVEL 26
- LEVEL 24
- STRUCTURAL STEEL OUTRIGGER TRUSS
- STRUCTURAL STEEL COMPOSITE FLOOR FRAMING TYPICAL
- REINFORCED CONCRETE MAT FOUNDATION WITH PILES
- FOUNDATION

TYPICAL HOTEL FRAMING PLAN

Labels:
- COMPOSITE WIDE-FLANGED GRAVITY FRAMING
- 8-COMPOSITE STEEL/ CONCRETE MEGA-COLUMN (CAST-IN-PLACE CONCRETE)
- CANTILEVERED CONCRETE JOIST FRAMING IN ATRIUM AREA
- COMPOSITE MEGA-COLUMN
- HOTEL ATRIUM
- 54 M
- CONCRETE CORE WALL W/ LINK BEAMS
- 8-STRUCTURAL STEEL BUILT-UP MEGA-COLUMN

TYPICAL OFFICE FRAMING PLAN

Labels:
- COMPOSITE WIDE-FLANGED GRAVITY FRAMING
- 8-COMPOSITE STEEL/ CONCRETE MEGA-COLUMN (CAST-IN-PLACE CONCRETE)
- COMPOSITE BUILT-UP FLOOR TRUSS GRAVITY FRAMING
- COMPOSITE MEGA-COLUMN
- 54 M
- CONCRETE CORE WALL W/ LINK BEAMS
- 8-STRUCTURAL STEEL BUILT-UP MEGA-COLUMN

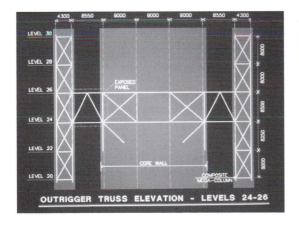

OUTRIGGER TRUSS ELEVATION - LEVELS 24-26

Labels: LEVEL 30, LEVEL 28, LEVEL 26, LEVEL 24, LEVEL 22, LEVEL 20; EXPOSED PANEL; CORE WALL; COMPOSITE MEGA-COLUMN
Dimensions: 4300 8550 9000 9000 9000 8550 4300; 8000 8000 8500 8250 8000

CLOCKWISE FROM TOP LEFT
Structural System Elevation,
Framing Plans,
Outrigger Truss Elevation,
Original Pin Concept Model
Jin Mao Tower, Shanghai, China

7.3.5   Pin-Fuse Devices

Based on the concept that structures should behave dynamically rather than statically during seismic events, the Pin-Fuse Joint® allows joints to be fixed in building frames until they are subjected to extreme loads. High-strength bolts, brass shims, and curved steel plates create fixity with a well-defined coefficient of friction. No movement occurs in joints during typical service conditions, including wind and moderate seismic events.

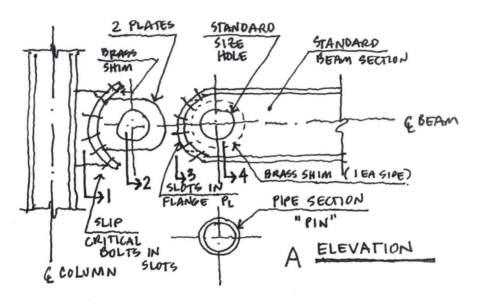

Original Concept Drawing, The Pin-Fuse Joint®,
Patent No. US 6,681,538 B1 & US 7,000,304

When an extreme earthquake subjects the structural frame to large forces and displacements, the moment-resisting beams rotate around central pins located in beam webs, breaking the coefficient of static friction and consequently dissipating energy. The frames act as mechanisms that soften the structure, lengthen the building period, and attract less force from the ground. Bolt tension is maintained so the coefficient of static friction is re-established between the curved plates and brass shim, resulting in complete pre-earthquake fixity without any inelastic deformations.

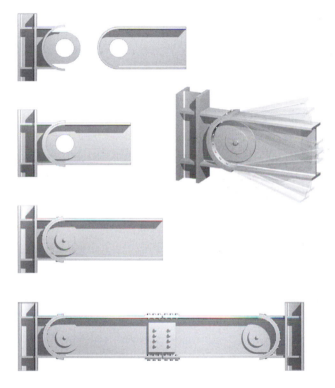

Renderings of the Pin-Fuse Joint®,
Patent No. US 6,681,538 B1 & US 7,000,304

CHAPTER 8
NATURE

IN 1956, FRANK LLOYD WRIGHT conceived of the Mile-High Building. Since Wright, others have followed with concepts that approach one mile in height. Foster and Partners have proposed the Millennium Tower and Cervera & Pioz have proposed the Bionic Tower. Realistic steps toward this extreme height are incremental, with what was formerly the tallest building in the world, Taipei 101, designed by C.Y. Lee and Partners, standing at 508 m (1667 ft). At 828 m (2716 ft), the Burj Khalifa, designed by SOM and completed in 2010, has significantly surpassed this height.

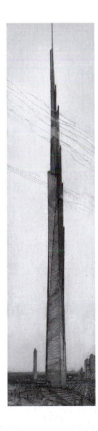

The Illinois:
The Mile High Building,
Frank Lloyd Wright

FACING PAGE
Bamboo in Nature

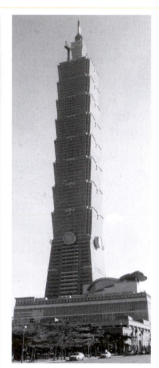

Millennium Tower,
Foster + Partners

Bionic Tower, Maria Rosa Cervera,
Javier Gómez Pioz

Taipei 101,
C.Y. Lee & Partners

Variations on current structural systems and the development of new systems are important for the next generations of ultra-tall buildings. Variations developed to solve structural challenges while achieving extreme height often lead to interesting architectural solutions as well.

8.1   SCREEN FRAMES

Rigid frames for buildings can be stiffened to create further resistance to wind and seismic loads. The screen frames used on the Goldfield International Garden, Beijing, use lateral stiffening frames within a mega-frame concept. Economical moment-resisting frame heights can be increased from 35 stories to 50 stories.

The Rigid Frame,
Inland Steel Building, Chicago, IL

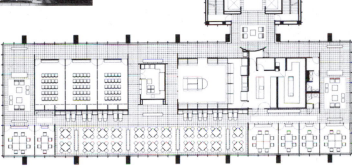

The Stiffened Screen Frame,
Goldfield International Garden, Beijing, China

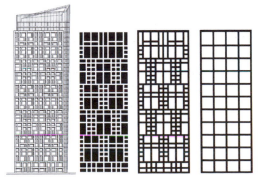

8.2   CORES AND PRESTRESSED FRAMES

The central core wall used to surround back-of-house areas, elevators, restrooms, and mechanical rooms is an excellent solution to tall building structures. For a building up to 40 stories, the proportion of core walls needed to surround the service areas naturally satisfies lateral and gravity requirements, with wall thicknesses typically not exceeding 600 mm (24 in). As heights are increased, these core walls typically increase in plan area (for instance, the number of elevators is greater); however, they are not sufficient to provide efficient resistance to drift. Therefore, combining these cores with perimeter frames through rigid diaphragms increases lateral resistance by sharing applied forces. Typically, the shear wall core dominates behavior at the base of the building, with the frames

Core-PT Frame,
500 West Monroe, Chicago, IL

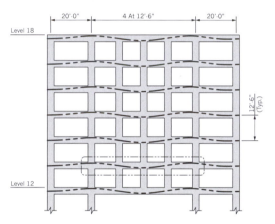

West Facade Transfer · Levels 12 · 18

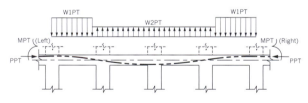

Effects of Post · Tensioning Only

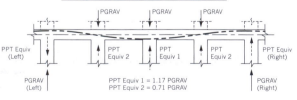

Load Balancing of Applied Gravity Loads

Core Wall,
NBC Tower at Cityfront Center, Chicago, IL

Typical Post-Tensioned Frame Details,
500 West Monroe, Chicago, IL

restricting the core wall cantilever behavior by acting to pull the core back at the top of the building. Managing the placement of gravity loads on frames can further increase the frame's efficiency by using gravity to counteract uplift forces. Post-tensioned concrete frames within the perimeter of 500 West Monroe in Chicago were used to redistribute axial loads, provide additional clear spans between columns, and increase structural efficiency.

8.3   THE INFINITY COLUMN

Dr. Fazlur Khan understood the limitation of conceiving a tall building as a tube, with solid but thin walls. The introduction of openings for windows was a must. However, he discovered that the placement and portioning of openings could still lead to an efficient structure. He transformed this idealistic concept into a constructable, affordable system by developing closely spaced columns and beams to form a rectilinear grid. Simply introducing diagonal members into the tubular frame achieved greater efficiency with height. This system was conceived for buildings consisting of all-steel and all-concrete.

The Jinao Tower in Nanjing, China, combines the use of local materials (reinforced concrete) and local labor to minimize costs for what would

Concrete Tube-in-Tube,
Chestnut-Dewitt Tower, Chicago, IL

Concrete Braced Tubular Frame,
Onterie Center, Chicago, IL

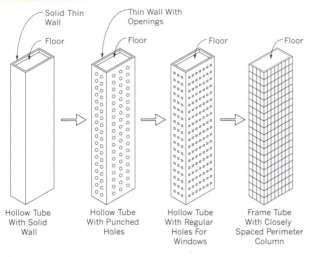

Dr. Fazlur Khan's Conceptual Tube Systems

The Infinity Column Concept,
Jinao Tower, Nanjing, China

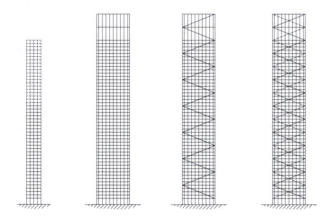

(left to right) Inner Tube Frame, Exterior Tube Frame,
Exterior Tube w/Diagonals, Exterior Tube w/Diagonals Beyond
The Infinity Column Concept,
Jinao Tower, Nanjing, China

otherwise be a conventional tube-in-tube structure. By introducing a diagonal steel member on each façade with a primary connection every four stories on lower floors and every five stories on upper floors, 45% of the rebar and concrete required for the lateral system could be eliminated, resulting in a 20% decrease in material overall.

8.4   GROWTH PATTERNS

The super-frame concept was developed for the initial designs of the 137 story Columbus Center in New York. A diagonal frame wrapped angular form was designed to confuse the wind. While SOM developed the competition scheme for the China World Trade Center, Beijing, interest in bamboo as an architectural form led to the discovery of the properties of bamboo and how they might relate to the extreme high-rise. The natural formation of bamboo reveals unique structural characteristics. Long, narrow bamboo stems provide support for large foliage during its growing life, while providing strong and predictable support for man-made structures after harvesting. Even when subjected to tsunamis, bamboo behaves effectively and efficiently in response to lateral loads, exhibiting the genius of natural structural properties and geometric proportioning. The nodes or diaphragms seen as rings over the height of the culm or stack are not evenly spaced—they are closer at the base, further apart through the mid-height, and close again near the top. These diaphragm locations are not random and can be predicted mathematically; they are positioned to prevent buckling of the thin bamboo walls when subjected to gravity and lateral loads. This growth pattern is common to all bamboo. The wall thicknesses and diameter of the culm can be similarly calculated. They are also proportioned to prevent buckling of the culm.

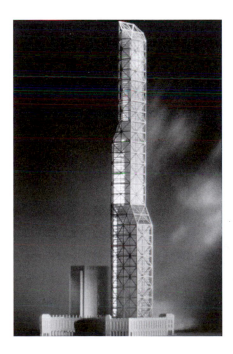

NY Coliseum at Columbus Center, Concept Model
New York, NY

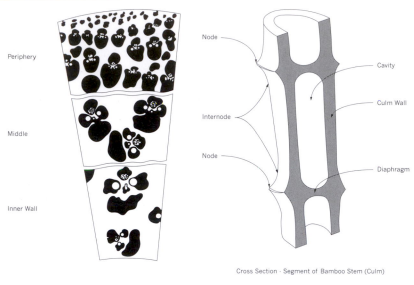

Bamboo Cross Section

Bamboo Culm

Cross Section · Segment of Bamboo Stem (Culm)

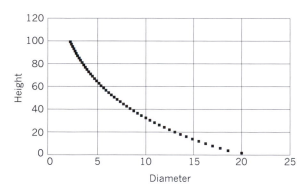

Bamboo Node Diameter vs. Height

Bamboo consists of a culm, or stem, comprised of nodes and internodes. Nodes mark the location of diaphragms and provide the location for new growth. A slight change in diameter exists at node locations. Internodes exist between nodes. Internodes are hollow, creating an inner cavity surrounded by a culm wall. Material in the culm is located at the farthest point from the stem's neutral axis, providing greatest bending resistance, allowing gravity loads to exist only in the outside skin which impedes uplift due to lateral loads and minimizes overall weight. The cellular structure of the bamboo wall reveals tighter cellular density near the outer surface of the wall and less density near the inner wall, again reinforcing the idea of maximum material efficiency when subjected to bending loads.

The geometric characteristics of bamboo are applied to the structural systems of the China World Trade Center Tower Competition submission. The tower is divided into eight segments along its height. The structural demand from lateral load is highest at the base of the culm (or tower), and therefore the internode heights are smaller compared to those at the mid-height. Smaller spacing increases moment capacity and buckling resistance. Beyond the mid-height of the culm (or tower) the heights of the internodes decrease proportionally with the diaphragm diameter. Thus, the form of the culm (tower) responds to structural demands due to lateral loads. The geometric relationships, such as length, culm diameter, and wall thickness, of many bamboo species have been previously discussed by Janssen (1991). Equations 1 through 4 define the bamboo form as discussed by Janssen (1991). Coefficients are an average of several cited species.

Internode number

$$x_n = n * \frac{100}{N} \tag{1}$$

Internode length

$$y_{n1} = 25.13 + 4.8080x_n - 0.0774x_n^2 \text{ (below mid-height)} \tag{2a}$$
$$y_{n2} = 178.84 - 2.3927x_n + 0.0068x_n^2 \text{ (above mid-height)} \tag{2b}$$

Internode diameter

$$d_{n1} = 97.5 - 0.212x_n + 0.016x_n^2 \text{ (below mid-height)} \tag{3a}$$
$$d_{n2} = 178.84 - 2.3927x_n + 0.0068x_n^2 \text{ (above mid-height)} \tag{3b}$$

Wall thickness

$$t = 35 + 0.0181(x_n - 35)^{1.9} \tag{4}$$

Here, x_n is the internode number; n is a shaping parameter specified by the architectural design team to be 80 based on the number of floors; N is the height of the structure (320 m); y_n is the internode length; d_n is the internode diameter; t is the wall thickness. For the internode length and diameter a non-linear relationship is observed by the transition from y_{n1} to y_{n2} and d_{n1} to d_{n2} at x_n. Thus, two polynomial equations are provided.

The relationships are shared among inner and outer structural systems. The outer structural system follows the internode length (Equation 2) with respect to mega-brace heights and mimics the culm wall fibers. The inner structural system also follows specified bamboo characteristics. Outriggers are taken as the "diaphragm" in bamboo, since outriggers tie perimeter structural systems in a similar manner to diaphragms in bamboo. Internode lengths are largest at mid-height and smallest at the base and

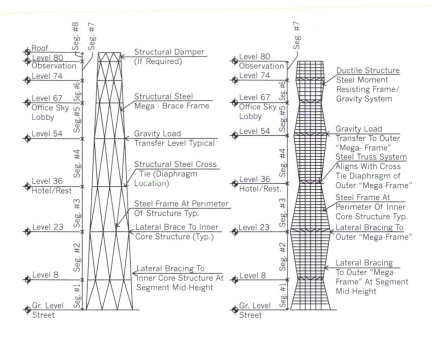

Seg. #8 Seg. #7

Roof
Level 80
Observation
Level 74

Level 67
Office Sky
Lobby

Seg. #6 Seg. #5 Seg. #4

Level 54

Level 36
Hotel/Rest.

Seg. #3

Level 23

Seg. #2

Level 8

Seg. #1

Gr. Level
Street

Structural Damper
(If Required)

Structural Steel
Mega · Brace Frame

Gravity Load
Transfer Level Typical

Structural Steel Cross
· Tie (Diaphragm
Location)

Steel Frame At Perimeter
Of Structure Typ.

Lateral Brace To Inner
Core Structure (Typ.)

Lateral Bracing To
Inner Core Structure At
Segment Mid-Height

Seg. #7

Level 80
Observation
Level 74

Level 67
Office Sky
Lobby

Seg. #6 Seg. #5 Seg. #4

Level 54

Level 36
Hotel/Rest.

Seg. #3

Level 23

Seg. #2

Level 8

Seg. #1

Gr. Level
Street

Ductile Structure
Steel Moment
Resisting Frame/
Gravity System

Gravity Load
Transfer To Outer
"Mega· Frame"
Steel Truss System
Aligns With Cross
Tie Diaphragm of
Outer "Mega-Frame"

Steel Frame At
Perimeter Of Inner
Core Structure Typ.

Lateral Bracing To
Outer "Mega-Frame"

Lateral Bracing
To Outer "Mega-
Frame" At Segment
Mid-Height

A Elevation-Outer
Mega-Brace Structural Frame

B Elevation-Inner Ductile
Moment-Resisting Frame

Bamboo Concepts—Structural System Elevations
China World Trade Center Competition, Beijing, China

Bamboo Concepts,
China World Trade Center Competition, Beijing, China

top. Diaphragm diameter is also varied over the height at outrigger levels, as specified by Equation 3. Finally, member sizes are proportioned to follow the wall thickness relationship shown in Equation 4.

All equations that define the diaphragm locations, diameter, and wall thickness are based on a quadratic formulation. When the required diameter of the culm is plotted against height (with the relationships of diaphragms and wall thicknesses similar), it mimics the bending loading diagram of a cantilever subjected to uniform lateral loads—the engineering theory is the same for bamboo and other cantilevered structures.

8.5   THE STAYED MAST

Interconnecting a central core with perimeter mega-columns or frames provides an excellent solution to the tall tower. Building use is maximized, with disruption to interior spaces being limited to local areas within the structure. Outrigger trusses or walls act as levers to prop central cores. The core can

Jin Mao Tower,
Shanghai, China

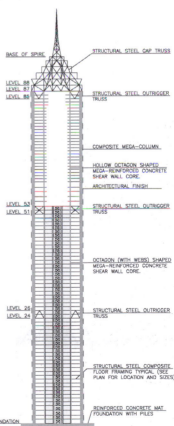

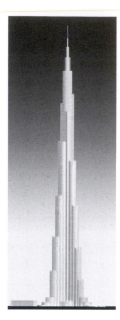

Burj Khalifa, Dubai, UAE

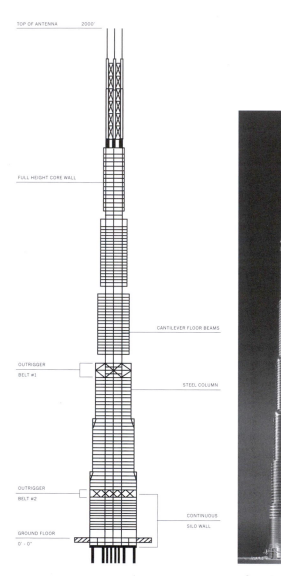

TOP OF ANTENNA   2000'

FULL HEIGHT CORE WALL

CANTILEVER FLOOR BEAMS

OUTRIGGER
BELT #1

STEEL COLUMN

OUTRIGGER
BELT #2

CONTINUOUS
SILO WALL

GROUND FLOOR
0' - 0"

7 South Dearborn, Chicago, IL

be considered a slender mast, stayed by the levers and perimeter columns. These levers develop the entire depth of the structure and are usually located at two or three locations within the height of the structure. Outrigger trusses or walls usually can be placed in mechanical areas within the tower. This system allows for heights of 365 m (1200 ft) and above. The Burj Khalifa, Dubai, completed in 2010, is the tallest building in the world.

8.6   THE PERFECT TUBE

Dr. Khan discovered that bundling tubular frames decreased shear lag. Shear lag is the inability of axial loads to flow around the cross-section of a tube when subjected to lateral load. Bundling the individual tubular frames not only decreases shear lag, but also increases efficiency. Efficiency is the ratio of axial column deformation to total deformation (axial, bending, and shear) in the tower. Khan sought solutions to the perfect tube and was able to increase the efficiency of the tubular frame from 61% to 78% (considering the geometry of the Sears Tower) by bundling the tubes, although he never achieved 100% efficiency.

The mesh tube conceived for the Jinling Hotel, Nanjing, China, incorporates a fine diagonal mesh of structure at the perimeter of the building, eliminating any significant local shear or bending deformations of vertical members, resulting in an essentially 100% efficient structure, a structure that might lead us to expanding further the height limits of the extreme high-rise.

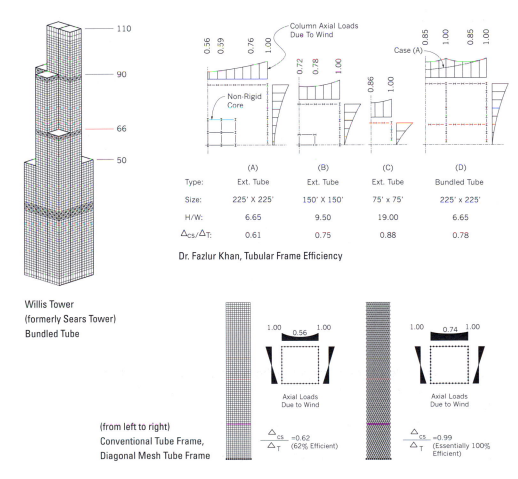

Dr. Fazlur Khan, Tubular Frame Efficiency

Type:	Ext. Tube	Ext. Tube	Ext. Tube	Bundled Tube
	(A)	(B)	(C)	(D)
Size:	225' X 225'	150' X 150'	75' x 75'	225' x 225'
H/W:	6.65	9.50	19.00	6.65
Δ_{cs}/Δ_T:	0.61	0.75	0.88	0.78

Willis Tower
(formerly Sears Tower)
Bundled Tube

(from left to right)
Conventional Tube Frame,
Diagonal Mesh Tube Frame

$\dfrac{\Delta_{cs}}{\Delta_T} = 0.62$ (62% Efficient)

$\dfrac{\Delta_{cs}}{\Delta_T} = 0.99$ (Essentially 100% Efficient)

8.7 THE LOGARITHMIC SPIRAL

The logarithmic spiral—found in forms ranging from shells, seeds, and plants to spider webs, hurricanes, and galaxies—can be interpreted and applied to ultra-tall towers, scientifically mimicking natural force flows of a cantilevered structure to its foundation.

The proposed design for the Transbay Transit Tower Competition in San Francisco, California, is inspired by natural forms. These mathematically derived forms define systems that are safe, sustainable, cost-effective to construct, and provide optimal performance in seismic events.

CLOCKWISE FROM LEFT
Weather System
Storksbill Seed
Spider Web
Nautilus Shell

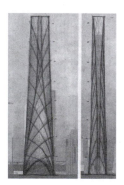
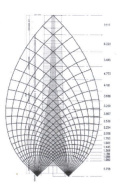
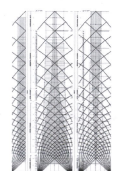
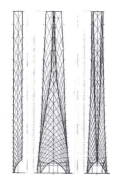

Michell Truss Diagram Applied to the Tower Elevations
Transbay Tower Competition, San Francisco, CA

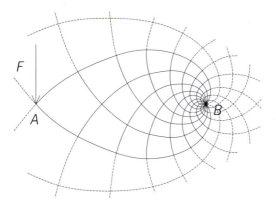

Michell Truss Diagram, A single force *F* is applied at *A*, and acting at right angles to line *AB*, is balanced by an equal and opposite force and a couple, of moment *F* x *AB*, applied at *B*. The minimum frame is formed of two similar equiangular spirals having their origin at *B* and intersection orthogonally at *A*, together with all other spirals orthogonal to these and enclosed between them

The spiral inherent in these natural forms traverses around a fixed center and gradually recedes from the center. Engineer Anthony Michell captured this behavior through his research in the early 1900s by describing the radiating lines of a pure cantilever, where force flow lines of equivalent constant stress result in specific spacing and orientations from the fixed support to the tip of the cantilever. The result is the most efficient cantilever system with the least material.

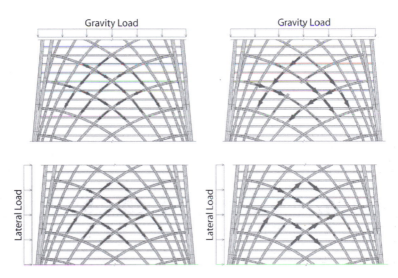

Alternative Load Path Force Flow Diagrams
Transbay Tower Competition, San Francisco, CA

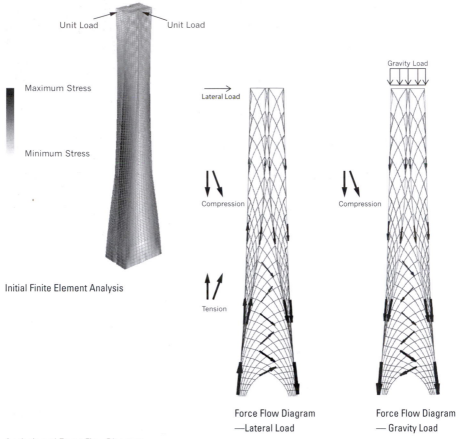

Initial Finite Element Analysis

Force Flow Diagram
—Lateral Load

Force Flow Diagram
— Gravity Load

Analysis and Force Flow Diagrams
Transbay Tower Competition, San Francisco, CA

The Michell Truss Diagram is mathematically interpreted and over-laid on the tower form, defining an optimal perimeter bracing configuration. The structure at the base of the tower is designed to accommodate a "gate-way" to the adjacent Terminal Building. The structural bracing responds to the openings in the structure, where demand is least, while mimicking grav-ity and lateral load force flows.

The exterior shell of the tower is robust by design, much like a spider web. It is able to self-heal in the unlikely event that a member is violated by fire or some other catastrophic event. Forces would flow to neighboring members down through the structure and ultimately into the foundations as shown. Should the structure's perimeter come under attack, the exterior composite (steel combined with concrete) bracing acts to disrupt or "shred" projectiles.

The "Perfect Tube,"
Jinling Hotel, Nanjing, China

CHAPTER 9
MECHANISMS

BEYOND LIFE SAFETY CONSIDERATIONS, increasing a building's service life is paramount in regions of high seismic risk. With the implementation of scientific structural devices and systems, building services, contents, and economic investments can be protected. Building codes specifying equivalent static analyses in many cases lead to conventional designs including members and joints that have limited ductility and have questionable economic value following a major seismic event.

What if structures were designed to behave dynamically, moving freely at times, dissipating energy, protecting life safety, protecting investments, and allowing structures to remain elastic after a severe earthquake, achieving the highest level of structural sustainability? What if these structures looked directly to nature for their mathematical derivations? What if the solutions to superior performance use conventional building materials? These solutions could provide a scientific response without great expense and construction complexity, while increasing a building structure's life cycle in regions of high seismic risk.

9.1 UNNATURAL BEHAVIOR

The 1995 Kobe Earthquake, or the Great Hanshin Earthquake as it has been formally named, was an important catalyst for considering building structures' long-term performance and life cycle. Structures in Kobe exhibited unnatural behavior; many exhibited poor ductility and many collapsed during strong ground shaking. Mid-height and ground story collapses of structures were common due to large changes in stiffness and partial height use of structural steel within reinforced concrete columns. Reinforced concrete structures lacked confinement of vertical reinforcing steel. Because of this unnatural performance of buildings, 107,000 buildings were damaged and

FACING PAGE
Transbay Tower Competition Rendering,
San Francisco, CA

Mid-height Collapse, Kobe, Japan (1995)

56,000 collapsed or were heavily damaged. Some 300,000 people were left initially homeless and 5000 people lost their lives. The total cost of the damage was well over $100 billion (1995 US dollars).

The solution to earthquake damage is to increase behavior predictability in very unpredictable events. The natural, unpredictable event of an earthquake cannot be changed; however, forcing the structural system to behave in a predictable manner can be achieved. The solution lies in the elasticity of materials, which can be closely predicted, and in the passive dissipation of energy along with the inherent, internal damping of structural systems.

9.2   CONVENTIONAL BEAM-TO-COLUMN TESTS

The mid-1980s brought the conclusion of many steel beam-to-column moment connection tests. These pre-Northridge studies investigated frame connections, including the fully welded top and bottom beam flanges and conventional shear tab web connections. The tests focused on these economical connections that incorporated typically available steel column and beam wide-flanged sections.

The tests compared the behavior of the connections without any column reinforcement to connections with flange continuity plates and connections with web doubler plates. Each connection was subjected to seven cycles of motion and was tested to at least 2% rotation to simulate expected rotation/drift in actual frame structures. Small rotations illustrated stress concentrations in beam and column flanges at complete penetration welded joints. Shear yielding in column webs could also be seen at small rotations. As larger rotations were imposed and connections were subjected to further cyclic motion, high material stress concentrations turned into fractures.

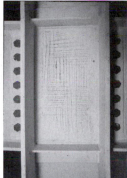

Beam-to-Column Joint Testing, Lehigh University, Bethlehem, PA

Typically, flange welds failed as a result of either failure of the weld itself or laminar tearing of the beam flange. Continuity plates provided better results, typically protecting column–beam flange connections; however, premature flange fractures were observed. Doubler plates protected the column webs but did little to reduce stress concentrations and eventual fracture of flange welds. These plates also severely limited the rotation capabilities of the joint. In some tests, flange welds were removed after fracture, re-welded, inspected and tested. These connections were further tested and in some cases met the required overall rotation and drift requirements. In all cases, connections generally remained intact; beams and columns did maintain connectivity through the bolted web connections and the other full penetration welds that remained after the initial fracture.

The 1994 Northridge Earthquake in southern California proved what was found through the connection testing in the mid-1980s. Steel moment frames experienced failure of welded flange connections but generally remained intact following the ground shaking. Life safety was protected; however, the economic loss was significant. Structural engineers were concerned about the effects of future events on frames with questionable connections. Retrofits were made through a very difficult process of gaining

Post-Northridge Reduced Beam Section (RBS) or "Dogbone" Connection

149

access to connections through building finishes, while owners wished to keep buildings occupied.

The reduced beam section or dogbone connection was developed as an economical solution to the joint while protecting the column and ensuring ductility.

9.3   WOOD DOWELS AND STEEL PINS

Jesuit churches built in the early 1900s on the Island of Chiloe off the coast of Chile have performed well over their life when subjected to strong ground motion. The 1960 Chilean earthquake, perhaps the strongest earthquake ever recorded at 9.5 on the Richter Scale, did not cause major damage to these structures. Witnesses of this event describe waveforms in large open fields with amplitudes over 6 feet. Street lights and poles used for telephone lines moved with large amplitudes and in some cases were observed to be almost parallel with the ground.

These churches have a common structural characteristic. Wood dowels from the indigenous Alerce tree were typically used to connect structural wood members with no "fixed joint" mechanical fasteners. Wood dowels were driven through openings in the structural wood members so that the dowels were snug tight and "fixed" during typical service conditions with an ability to rotate during extreme loading conditions. When the joints rotated, energy was dissipated, with the structures softening. With this softening the structures' period lengthened and as a result attracted less force from the ground motion.

(top) Wood Dowels

(left) Chilean Church Using Wood Dowel Connection Joints

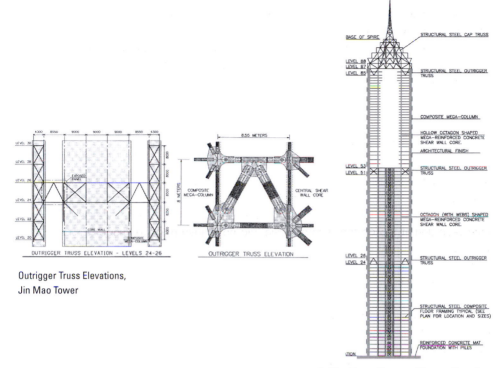

Outrigger Truss Elevations,
Jin Mao Tower

Jin Mao Tower Structural System Elevation

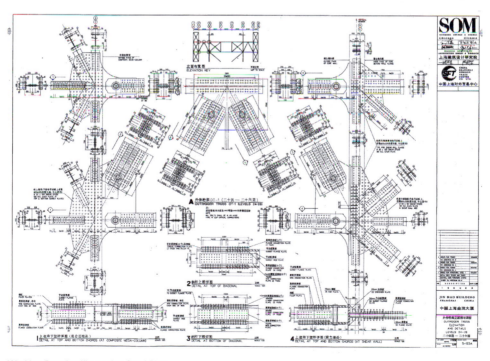

Working Drawing Illustrating Steel Pins,
Jin Mao Tower, Shanghai, China

Large diameter steel pins proved useful in the design of key structural elements of the Jin Mao Tower, Shanghai, in the People's Republic of China. The structural system concept for the 421 m (1380 ft) tall, 88-story tower included lateral load-resisting outrigger trusses that interconnected the central reinforced concrete core and eight perimeter composite mega-columns. This interconnection occurred at three two-story segments between Levels 24 and 26, 51 and 53, and 85 and 87. Relative creep, shrinkage, and elastic shortening between the core and the composite mega-columns was significant; if this behavior was not managed, the forces developed in the outrigger trusses would be large enough to overload the members, causing significant oversizing, yielding, and potential failure. If fully connected at the time of erection, up to 45 mm (1.75 in) of relative displacement was expected between the core and composite columns, with only 16 mm (0.625 in) of relative displacement expected after 120 days. By delaying the fully bolted connections for 120 days and including the pins to allow for free movement, the structure was only required to accommodate 16 mm (0.625 in) of relative displacement after the bolts were installed. To manage these forces and reduce the amount of structural material required for the members and connections, large-diameter steel pins were incorporated into the truss system. Based on basic concepts of statics, these pins allow the trusses to behave as freely moving mechanisms prior to making final connections. This mechanism concept allowed the trusses to be constructed during normal steel erection procedures and allowed final bolted connections to be installed after a significant amount of relative movement had already taken place. Because diagonal members were used within the trusses, large slots were incorporated into end connections to allow movement to occur. These slots were the key component to all free movement of the system.

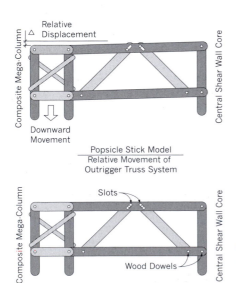

Popsicle Stick Model
Relative Movement of
Outrigger Truss System

Outrigger Truss System
Popsicle Stick Model,
Jin Mao Tower, Shanghai, China

A simple model made up of tongue depressors, wood dowels, and popsicle sticks was used to study the behavior. This conceptual model was the basis for the structural solution considering building materials, then analyzed and developed into full working drawings.

9.4 PINNED JOINTS

The Jin Mao pinned-truss concept, based on the fundamentals of statics and behavior, led to the development of a series of structural systems that will behave predictably in an extreme seismic event. The systems are appropriate for buildings and other structures with varying heights and geometries, allowing systems to remain elastic and to dissipate energy while protecting economic investments.

Structural steel end connections of members contain pins or bolts and are installed to a well-calibrated torque by applying compression of joints through the pin or bolt tension. Faying surfaces are treated with an unsophisticated slip-type material such as brass, bronze, cast iron, aluminum or hard composite materials with a well-defined coefficient of friction, allowing for significant movement capabilities after the threshold of slip and without significant

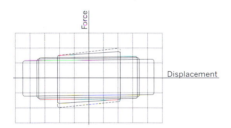

Representative Hysteresis Loop

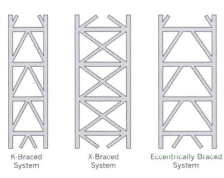

K-Braced System   X-Braced System   Eccentrically Braced System

Truss System Elevations

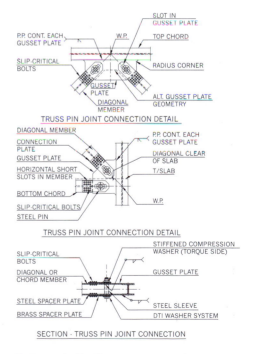

Pin-Connection Details

loss of bolt or pin tension. The material combination not only provides well-defined load-displacement characteristics, but also excellent cyclic behavioral attributes. Structural steel surfaces can be sandblasted or cleaned with mill scale present to minimize galling or binding. A force vs. displacement hysteresis curve of a combination of steel and brass illustrates the stable behavior.

Pins, installed into the end connections of trussed building frames used in buildings subjected to extreme seismic events, provide fixed joints for the typical service life of the structure and are allowed to slip during extreme events.

Circular holes are used in horizontal truss members for both connection plates and members. Circular holes are used in diagonal members with long-slotted holes used in connection plates. Slots are designed and installed in the direction of force. The length of the hole is dependent on the predicted drift and the dynamic characteristics of the structure when subjected to ground motion. A well-defined faying surface is used between connection plates and members. The material sandwiched between the steel surfaces is a shim or pad form (may be applied directly to the steel surfaces) and is stringently protected after application, during shipment, and during erection. The faying surface material is used in both slotted and circular connection joints. Pins are torqued in slip-defined joints at slotted connections of diagonal members. The size of the pin and the amount of torque are directly related to the coefficient of friction of the material being compressed between the steel plates and the expected force that will initiate joint motion. Special Belleville washers are used to help maintain bolt tension after slip has occurred in slotted joints. In addition, direct tension indicators (DTIs) are used under the head (non-torquing side) to ensure proper pin or bolt tension. Structural members are designed for forces required to initiate the onset of joint motion. Members are designed to remain elastic, not allowing yielding or local or global buckling.

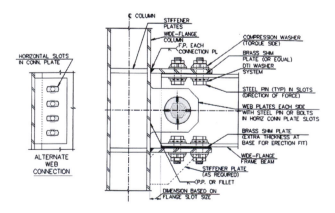

Moment Frame Beam-Column Joint Detail

Period vs. Total Number of Slipped Members

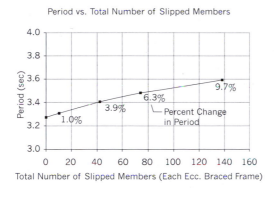

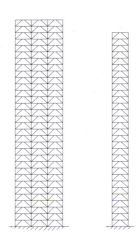

Period vs. Total Number of Slipped Members, Truss Example

Moment-resisting frames equally benefit when pins are introduced into end connections. Pins in long-slotted connections can be introduced into top and bottom beam flanges. The length of the beam flange slots can be predicted by evaluating the expected beam rotation and inter-story drift characteristics. A single pin in a circular hole is used in the web connection to provide the best rotational capabilities.

These connections, whether placed in braced or moment frame systems, act to lengthen the structure's period. This lengthening of the period effectively softens the structure, resulting in less force being attracted to the structure from the ground motion.

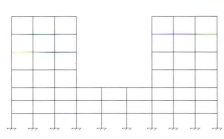

Period vs. Total Number of Slipped Members, Frame Example

Period vs. Total Number of Slipped Members

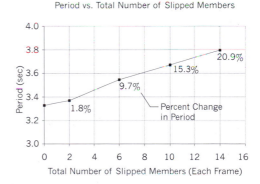

9.5   THE PIN-FUSE JOINT®

If structures were capable of altering their characteristics to resist potentially destructive forces during extreme loading events without permanent deformation, their expected life cycles would significantly increase.

Structural frames previously utilizing beam-to-column moment connections that are welded with the frame beams perpendicular to the columns would become obsolete. Beams are typically connected to the face of columns and rotate when subjected to racking of the building frame. These beams are designed to protect the column integrity and prevent potential collapse by plastically deforming during frame motion. This deformation, however, likely decreases post-earthquake integrity and economic viability in the process. Therefore, a new concept for this connection is warranted.

The Pin-Fuse Joint® allows building movement caused by a seismic event, while maintaining structural elasticity after strong ground motion. The joint introduces a circular-plated end connection for the steel beam framing into the steel or composite columns within a moment-resisting frame. Slip-critical friction-type bolts connect the curved steel end plates. A steel pin or hollow steel pipe in the center of the moment-frame beam provides a well-defined rotation point. Under typical service conditions, including wind and moderate seismic events, the joint remains fixed where applied forces do not overcome the friction resistance provided between the curved end plates. However, during an extreme event, the joint is designed to rotate around the pin, with the slip-critical bolts sliding in long-slotted holes in the curved end plates. With this slip, rotation is allowed, energy dissipates, and "fusing" occurs.

The rotation of the Pin-Fuse Joint® during extreme seismic events occurs sequentially in designated locations within the frame. As the slip occurs, the building frame is softened. The dynamic characteristics of the frame are altered with a lengthening of the building period, and less force is attracted by the frame; however, no inelastic deformation is realized. After the seismic event, the elastic frame finds its pre-earthquake, natural-centered position. The brass shim located between the curved steel plates provides a predictable coefficient of friction required to determine the onset of slip and also enables the bolts to maintain their tension and consequently apply the clamping force after the earthquake has subsided.

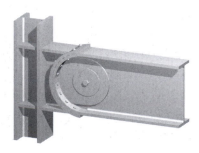

The Pin-Fuse Joint®,
Patent No. US 6,681,538 B1 & US 7,000,304

9.6   MANAGING LARGE SEISMIC MOVEMENTS

The use of pinned mechanisms in structures is not limited to frame joints; such mechanisms may also be used in large structural components that assist in natural building behavior during extreme seismic events and allow for increased building life cycles.

A rocker mechanism, installed at the mid-height of the New Beijing Poly Plaza, allowed the building to move freely during a significant earthquake, while providing support for the world's largest cable net and for a mid-height museum used to display some of China's most important antiquities. The cable net, 90 m (295 ft) tall and 60 m (197 ft) wide at its widest point, was conceived to provide support for the exterior wall system with minimal structure. Stainless steel cables were typically spaced at 1500 mm (59 in) on-center, with vertical cables of 26 mm (1 in) in diameter and horizontal cables of 34 mm (1.34 in) in diameter. A V-shaped cable stayed concept was used to reduce the cable net span and the overall displacements when subjected to wind loads. Cables 200 mm (8 in) in diameter not only provided lateral support for the cable net but also were used to support the mid-height museum. It was found that these large cables acted as diagonal mega-braces when the building was subjected to strong ground motion. The displacement at the top of the building relative to the mid-height was approximately 900 mm (36 in). The forces developed in the cables and connections with this level of relative displacement could not be reasonably resisted. If forces in the primary cables could be relieved during lateral motion, then cables and connections could be designed for wind and gravity loads only. A pulley, located at the top of the museum, could achieve this behavior. However, because a drum nearly 6 m (20 ft) in diameter was required to accommodate the primary cables and was aesthetically unacceptable, an alternate idea was required.

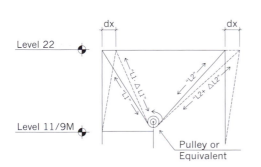

Final Scheme · Pulley Equivalent at Base of "V"

Pulley Concept,
New Beijing Poly Plaza, Beijing, China

The Rocker,
New Beijing Poly Plaza, Beijing, China

Study Model,
New Beijing Poly Plaza, Beijing, China

A rocker mechanism, capable of movements in two primary directions, was introduced into the top of the museum. Pins combined with steel castings provided the reverse pulley mechanism capable of resisting the imposed loads while remaining elastic during an extreme seismic event.

9.7 COMBINING NATURAL FORMS AND MECHANISMS

The combination of natural forms and mechanisms results in the minimum use of material and the best life cycle. Therefore, a dual system could be incorporated into a structure: one system controlling lateral drift and one providing fused mechanisms used to protect the structure while maintaining permanent elasticity, as illustrated in the plan for the Transbay Tower.

Transbay Tower Rendering

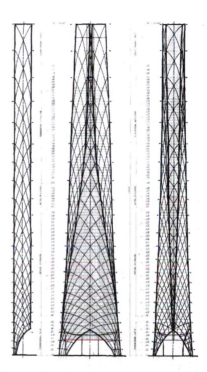

Transbay Tower Elevations

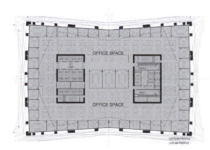

Transbay Tower Competition Floor Plan

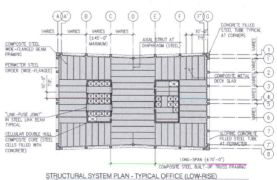

STRUCTURAL SYSTEM PLAN - TYPICAL OFFICE (LOW-RISE)

Transbay Tower Competition Structural System Plan

Link-Fuse Joint™ US Patent No. 7,647,734,
At-rest Shape

Link-Fuse Joint™ US Patent No. 7,647,734,
Displaced Shape

The core of the tower is designed to protect inhabitants and guarantee a safe path of egress in the event of an emergency. The core walls are hardened with a cellular structural concept derived from maritime construction. This "double hull" wall includes a steel plate shell filled with concrete and provides an armored barrier for elevators, stairs, and primary mechanical life safety systems. This barrier also provides optimal fire resistance.

The tower is designed to resist the most extreme earthquakes and remain operational in tandem with the essential Terminal facility. The tower incorporates an innovative array of seismic fuses designed to slide during extreme seismic events. This Link Fuse™ system allows the building core to dissipate energy at wall link locations (where openings are required to enter the core) while protecting the rest of the structure from damage. After the earthquake subsides and the building comes to rest, the fuses maintain their load carrying capacity and the building can immediately be put back into service.

The Link-Fuse Joint™, incorporated into the reinforced concrete shear wall (or steel frames) of structures, allows links that typically occur over doorways and mechanical openings to fuse during extreme seismic events. The butterfly slot pattern in the steel connection plates is clamped together, developing static friction between plates. Brass shims located between the plates create a well-defined and consistent coefficient of friction. When the demand on the links is extreme, the joints slip in any necessary vertical or horizontal direction, dissipating energy, softening the building (period lengthens), and attracting less force from the ground. After the motion of the building and the movement of the joints cease, the building and joints return to their natural at-rest position without permanent deformation. The tension in bolts used to clamp the plates together is not lost during the motion and, therefore, re-establishes the structural capacity by restoring the static friction within the joints. The structure is safe, its economic investment protected, and it can remain in service. In contrast, conventional reinforced concrete link beams are typically damaged and must be repaired or replaced, with the overall structure potentially being deemed unfit for future service.

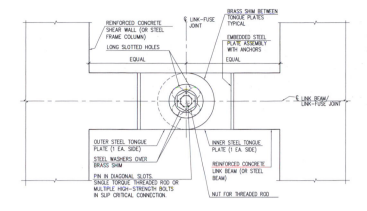

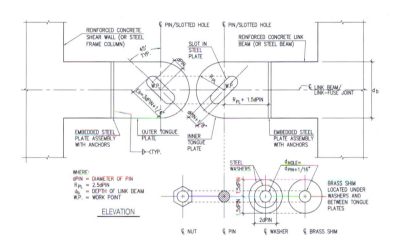

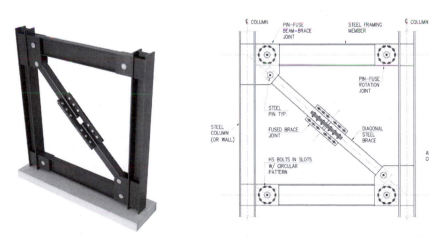

The Pin-Fuse Frame™, US Patent No. 7,712,266

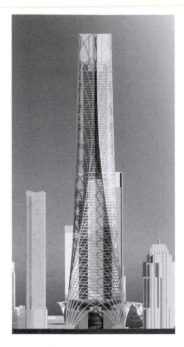

Transbay Tower Competition, San Francisco, CA

The Pin-Fuse Frame™, incorporated into braced frames (or between concrete shear walls) of structures, allows braces to slip or "fuse" during extreme seismic events. High-strength bolts in long-slotted holes are used to clamp a sandwich of brass and steel plates (brass between the plates). Brass placed between steel plates creates a well-defined threshold of slip when subjected to load. When the demand on the structure is extremely high, the moment-resisting frame with its circular bolt pattern provides additional resistance. If these joints are subjected to a high level of bending moment, they too will slip, rotating around a center steel pin. The combination in behavior of the braces and horizontal moment-resisting elements fuses the structure, dissipates energy, softens the building, lengthens the period, and reduces forces attracted from the ground motion. After the event, these joints return to their full structural capacity without permanent deformation. The tension in bolts used to clamp the plates together is not lost during the motion and, therefore, re-establishes the structural capacity by restoring the static friction within the joints. Although conventional steel braced frames have been proven to perform with some adequacy during extreme seismic events, their integrity is questionable in repetitive cyclic motions due to premature buckling. Damage in many cases may not be repairable.

Transbay Tower Competition, San Francisco, CA

The structural topology conceived at the perimeter combined with the fused core could provide what may be the most efficient structural system for a tall building anywhere in the world, reducing the material quantities required for construction and providing a truly sustainable structure designed not only to survive but also to remain in service after even the most significant natural and unnatural disasters.

CHAPTER 10
ENVIRONMENT

THE AVERAGE TEMPERATURE RISE on the planet between the start and the end of the 20th century has been measured to be almost 1.5°F. The Intergovernmental Panel on Climate Change (IPCC) has concluded that most of the observed temperature rise is due to increased amounts of greenhouse gases, with carbon dioxide being the major contributor. The most significant contributors to greenhouse gases resulted from human activities, including the burning of fossil fuels and deforestation. The industrial revolution in the United States and Europe in the early 1900s may have been the period of greatest significance. The industrial developments in China and India may take their place as the most significant contributor in the 21st century. Another 2°F temperature rise is expected to occur in this century (IPCC, 2007).

Global temperature increases will result in sea level rise, will change patterns of precipitation, and will perhaps expand subtropical deserts. The Arctic region is expected to see the greatest effects, with the retraction of glaciers and a reduction in permafrost and sea ice resulting in extreme weather patterns, the extinction of species, and changes in agricultural yields. Water level rises will have catastrophic effects on developments and ecosystems that currently exist along shorelines and waterways that interconnect with oceans.

Buildings account for 39% of the carbon emissions in the United States—more than either the transportation (33%) or industrial (29%) sectors. Most of the emissions are associated with the combustion of fossil fuels. The carbon emissions associated with manufacturing and transporting building construction materials as well as demolition or repair have an even greater impact (USGBC, 2004).

FACING PAGE
Post-consumer Waste Plastics

structures, where gravity loads are very large, uplift conditions could exist, and the period of the superstructure is typically long—sometimes longer than the achievable period of an isolation system.

When structures are fixed to their foundations, this movement must be designed to occur within the joints of the superstructure. Pin-Fuse seismic systems are designed to maintain joint fixity throughout the typical service life of the structure. When a significant seismic event occurs, forces within the frame cause slip in joints through friction-type connections. This slippage alters the characteristics of the structure, lengthening the period and reducing the forces attracted from the ground, and provides energy dissipation without permanent deformation.

10.3   REDUCTION OF SEISMIC MASS

The most efficient and environmentally responsible structures are those with least mass that incorporate structural solutions informed by nature. Reducing seismic mass can be accomplished through the use of light-weight materials such as light-weight concrete that has 25% less mass than normal-weight concrete. However, other concepts can be introduced that further reduce mass. In all concrete structures there are areas that include significant amounts of concrete only because of conventional construction practices. For instance, the concrete needed in the middle-middle strip of a two-way reinforced concrete floor framing system could be reduced by 50% or more by introducing more scientific systems. If concrete in these areas could be displaced where it is not required, this reduction could be achieved. An inclusion system, perhaps one that includes post-consumer waste products, could be used. The patent pending Sustainable Form Inclusion System™ (SFIS) accomplishes this both by creating voids within the concrete framing system and by using materials such as plastic water bottles, plastic bags, waste Styrofoam, etc. that would otherwise be placed in landfills.

Post-Consumer Plastics Formed
Into Rectangular Units

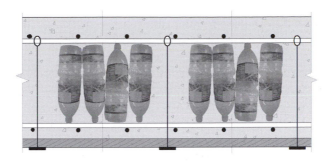

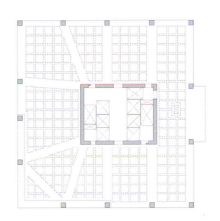

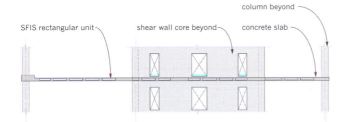

SFIS rectangular unit

shear wall core beyond

column beyond

concrete slab

CLOCKWISE FROM TOP

Sustainable Form-Inclusion System™ (Patent Pending)

Typical Framing Plan with SFIS Rectangular Units
in a High-Rise Structure

Typical Overall Section with SFIS Rectangular Units
in High-Rise Structure

Typical Reinforcing Mock-up with Water Bottles Used
in SFIS System for High-Rise Structure

10.4   THE ENVIRONMENTAL ANALYSIS TOOL™

10.4.1   The Basis of Evaluation

Up until now, most, if not all, of the efforts made in calculating the carbon footprint are associated with the operations of buildings, with little or no focus on the structure at the time of construction and over its service life. The Environmental Analysis Tool™ (Patent Pending) calculates the expected carbon footprint of the structure at the time of construction considering its location and site conditions. Based on the structural system, a complex damage assessment is performed accounting for the expected seismic conditions and anticipated service life, whether the structure is comprised of a code-defined conventional system or enhanced structural system.

10.4.2   Carbon Mapping Early in Design

The environmental impacts of structures need to be considered just as much as available materials, constructability, and cost at the earliest of stages of design. It is important that the carbon footprint assessment is accurate even with a limited amount of known information. The Environmental Analysis Tool™ is capable of calculating the structure's carbon footprint even when only the following information is known:

1. The number of stories (superstructure and basement).
2. The total framed area in the structure or average area per floor.
3. The structural system type.
4. The expected design life.
5. Site conditions related to expected wind and seismic forces.

With this limited amount of data, the program references to a comprehensive database containing the material quantities for hundreds of previously designed SOM structures, although any credible database can be used to define expected materials in the structure. Curve fitting techniques are used to consider building height relative to low, moderate or high wind/seismic conditions. Superstructure materials include structural steel, reinforced concrete, composite (combination of steel and concrete), wood, masonry, and light metal framing. Foundation materials include reinforced concrete for spread/continuous footings, mats, and pile-supported mats.

The selected seismic resisting system is important to the carbon footprint over the life of the structure. The contribution of carbon related to damage from a seismic event could account for 25% or more of the total carbon footprint for the structure. The basis of design includes conventional code-based systems with the option to select enhanced seismic systems such as the Pin-Fuse Seismic Systems, seismic isolation, unbonded braces, and viscous dampers. The repair required for damage associated with each system

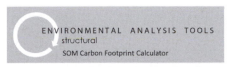

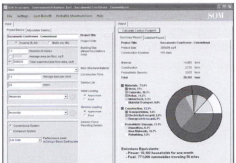

Environmental Impact Analysis

Sacramento Criminal Courthouse

Parameters:

- Steel (MF + BRBF)
- Conventional seismic system
- Steel Quantity: 16.5psf
- 396,609 sq. ft.
- 13 stories
- 25 yr life-cycle
- Seismic performance level:
  "Life Safety"

Estimated Carbon Footprint:

- Material   = 14,900 tons CO2
- Construction   = 2,730
  Seismic Damage = 2,620
  ➢ TOTAL   = 20,300 tons CO2

20,300 tons

Power: 10,100 households for one month
Fuel: 771,000 automobiles traveling 50 miles

Environmental Impact Analysis

Sacramento Criminal Courthouse

Parameters:

- Steel (MF + BRBF + Base Isolation)
- Enhanced seismic system
- Steel Quantity: 15.5psf
- 396,609 sq. ft.
- 13 stories
- 25 yr life-cycle
- Seismic performance level:
  "Operational"

Estimated Carbon Footprint:

- Material   = 14,500 tons CO2
- Construction   = 2,730
  Seismic Damage = 35
  ➢ TOTAL   = 17,200 tons CO2

17,200 tons

Δ = 3,100
tons

Power: 8,620 households for one month
Fuel: 655,000 automobiles traveling 50 miles
15% reduction

(top) Carbon Footprint Calculation without Considering Seismic Isolation (Patent Pending)

(bottom) Carbon Footprint Calculation Considering Seismic Isolation (Patent Pending)

is considered. The program uses this fundamental information to calculate the time and methods of construction, the fabrication and transportation of material, the labor required to build the structure, and laborers' transportation needs, among other things. Using this limited amount of information, an early but accurate assessment of the structure's carbon footprint can be made.

10.4.3   Carbon Mapping in Advanced Stages of Design

Every input variable in the Environmental Analysis Tool™ can be overridden by the user. This allows specific design information to be included in the evaluation after more comprehensive engineering is complete. For instance, specific structural steel, concrete, and rebar quantities can be used to accurately define the carbon footprint associated with materials. When specific supply locations for materials are known, the transportation distances can be used. Construction time can be modified based on anticipated complexities.

Default fragility curves that anticipate damage can be modified to accurately reflect the non-linear behavior of the structural system. Finally, the user can override specific assumptions made for carbon emission equivalents of anything from materials to transportation to construction.

10.4.4   Environmental Analysis Tool™ Program Details

Equivalent carbon dioxide emissions associated with the structural system of a building may be categorized as those resulting from the following three major components: (1) materials used to manufacture the structure; (2) construction; and (3) probabilistic damage due to seismicity. For instance, the primary material in a steel superstructure includes structural steel, concrete, rebar and metal deck, with manufacturing and transportation of the material from the source to the project site also being considered. Construction includes transportation of materials on the site, transportation of workers to and from the site, electricity required from the grid, and energy required from off-grid sources including such items as on-site diesel-powered generation. Finally, probabilistic damage due to seismicity includes demolition of damaged areas, repair, and replacement of structural components.

The carbon calculator considers the measurement of *equivalent* carbon dioxide emissions for a building structure. "Equivalent" is used to account for other gases besides carbon dioxide (CO_2) that are considered to be greenhouse gases and contribute to the total, 100-year global warming potential (GWP) of the structure in question, which is given in units of CO2e, or equivalent carbon dioxide. To sum up the contributions from each of these gases to the total GWP, factors are assigned to each gas based on molecular weight, using carbon dioxide as the benchmark. An example is methane; its GWP is 21 in equating it to CO2.

10.4.5   Cost–Benefit and PML

One of the biggest obstacles to introducing advanced engineering components into structures is first cost. Many of these systems will require a higher initial investment; however, when considered over the life cycle of the structure, cost-benefit and Probable Maximum Loss (PML) are extremely important.

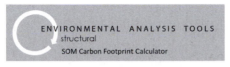

**Enhanced system saves $1,150,000 per year
on average versus conventional system**

Cost-Benefit Analysis Considering Seismic Isolation (Patent Pending)

Cost-Benefit Analysis

Sacramento Criminal Courthouse

Parameters:

- Base Building Cost: $ 438.6 million
- Enhanced System
  Estimated First Cost: $ 5.0 (1.1%)

Return on Investment:

- Expected Annual
  Loss Benefit   = $1,150,000 / yr
  ➤ ROI   = 23% over 25yrs

Benefit/Cost Ratio, 100-yr event:

- Reduction in loss = $13.6 million
  ➤ B/C Ratio   = 2.7

Benefit/Cost Ratio, 1000-yr event:

- Reduction in loss = $120.0 million
  ➤ B/C Ratio   = 24

The Environmental Analysis Tool™ considers the first cost of these systems and performs an analysis of anticipated damage and costs over the structure's specified life to calculate the cost-benefit ratios. The cost-benefit analysis considers the annual rate of return, mean annual loss savings, and first costs.

In addition, cost-benefit ratios are calculated for 100 and 1000 year seismic events. A cost-benefit ratio greater than 1 indicates a profitable investment. For instance, the state of California considers a minimum cost-benefit ratio of 1.5 to be required for the potential of a 100 year seismic event over a service life of 25 years.

Most insurance companies require a PML analysis. This analysis represents the total expected loss due to damage as a percentage of total cost of the building (including all components). The higher the PML, the greater the damage and the greater the expected cost to repair the damage. For code-compliant buildings the PML can be expected to range from 10 to 20; older structures designed to previous codes may have a PML of 20 or above; buildings with enhanced seismic systems, such as seismic isolation, will likely have a PML of 10 or less with values as low as 2 to 4.

10.5 REDUCING ENVIRONMENTAL IMPACT
THROUGH ADVANCED ORGANIC THEORIES

10.5.1 Emergence Theory

Structures that respect the growth patterns in nature will ultimately lead to minimal material use and effect on the environment. In many cases these forms are complex and require interpretation. Advanced computational and drawing tools have contributed to significant progress in developing these complex concepts. Such ideas are further defined by least energy principles related to strain energy and emergence theory concepts.

Intrinsic rules and relationships shape how elementary components can comprise complex organisms and systems. These rules and relationships often orchestrate the growth of the higher-level system without global oversight or guidance. Scientists have observed that these systems are organized, stable and complex. Emergence is a theme observed in nature

CLOCKWISE FROM TOP LEFT
Termite Colony

Bird Wing Skeleton Section

Honeycomb

Initial Form   Emergence Theory   Final Structural   Final Architectural
                         Diagram   Frame   Elevation

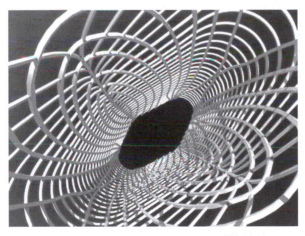

Final Perimeter Structural Frame,
Gemdale Tower Competition, Shenzhen, China

which suggests that complex, organized, and stable organisms and systems arise from relatively simple sub-components and their interactions without external guidance.

An example of an emergent system is the ant colony. Although the queen ant is the only member creating offspring, she does not direct the ant colony as a whole. The success of the colony is entirely dependent on basic relationships between ants without supervisory guidance, yet they are known to be highly organized. Also, the structures that ants build are the result of an emergent process, as they are built based on basic rules and regulations, but without a master plan. In spite of this, termites can build very tall and stable structures with self-cooling characteristics.

Another example of an emergent system is the bone structure of a bird's wing. Bird bones need to be light and structurally efficient to improve the capabilities of the host bird. Over time, bird wings have developed light-weight truss systems inside the wing bones. Bone framework is the result of small, unguided changes over time, and therefore is an example of emergence. A third example of emergence is the honeycomb. Bees build individual hexagonal compartments based on instinct, and a highly efficient structural and storage system emerges.

The overall premise of the concept is that the collaboration of a collection of individual elements has far greater strength than if individual elements act alone.

These same principles can be applied to structures of various forms and boundary conditions. Using the emergence theory and energy principles, the resulting structures require the minimum amount of material and have the smallest possible carbon footprint.

Emergence theory and the use of optimization through strain energy principles led to a perimeter structural steel frame containing 25% less structural steel when compared to a conventional, rectilinear steel frame for the Gemdale Tower Competition structural system. This reduction in structural steel along with the introduction of enhanced seismic systems such as the Pin-Fuse Frame™ resulted in a 30% reduction of carbon in this 71 story, 130,000 square meter (1.4 million square feet), 350 m (1150 ft) tall tower in Shenzhen.

10.5.2   Fibonacci Sequence

Some would argue that the mathematics used to define the Fibonacci Sequence may be the numerical definition of life. Galaxies, hurricanes, hair growth, plant growth, and the proportions of the human body are all based on the same sequential principle. These natural forms, based on the binary numbers 0 and 1, represent the most efficient structures.

In 1937 MIT student Claude Shannon recognized that the *and*, *or*, or *not* switching logic of Boolean Algebra developed almost one hundred years earlier was similar to electrical circuits. This yes–no, on–off approach to defining operations described in Shannon's thesis led to the practical use of binary code in computing, electrical circuits, and other applications. This was the breakthrough to Gottfried Wilhelm Leibniz's 17th century search for converting verbal logic statements to mathematical ones. The ones and zeros can be assembled into binary strings where 8 digits can be assembled to represent 256 possible combinations of different letters, symbols, or instructions.

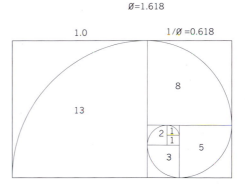

Fibonacci Sequence Inspiration,
Wuxi Times Square, Wuxi, China

Fibonacci Sequence/Golden Spiral

The binary digits found in these logic formations are prevalent in many other numerically defined forms, particularly those seen in nature. Those natural forms highlight organization, efficiency, and proportion. The golden spiral, mathematically defined by the Fibonacci Sequence, is rooted in logic formations created by binary numbers. The Fibonacci Sequence is a series of numbers where the subsequent number in the sequence is equal to the sum of the previous two numbers, starting with 0 and 1. The sequence of numbers is 0, 1, 1, 2, 3, 5, 8, 13, 21, 34, 55, 89, 144, 233, 377 ... More interesting is the multiplicative relationship between any two numbers, as each number approximates the previous number by the golden section multiple. This numerical ratio or proportional constant (Ø) converges on the divine limit as the numbers in the sequence grow. The value of this constant converges on 1.618.

The mathematical definition of the sequence is as follows:

$$F_n = F_{n-1} + F_{n-2}$$

where,

F = Fibonacci number
n = number in the sequence

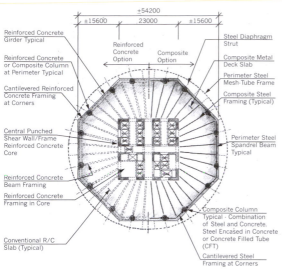

Reinforced Concrete Girder Typical

Reinforced Concrete or Composite Column at Perimeter Typical

Cantilevered Reinforced Concrete Framing at Corners

Central Punched Shear Wall/Frame Reinforced Concrete Core

Reinforced Concrete Beam Framing

Reinforced Concrete Framing in Core

Conventional R/C Slab (Typical)

+54200

±15600   23000   ±15600

Reinforced Concrete Option   Composite Option

Steel Diaphragm Strut

Composite Metal Deck Slab

Perimeter Steel Mesh-Tube Frame

Composite Steel Framing (Typical)

Perimeter Steel Spandrel Beam Typical

Composite Column Typical - Combination of Steel and Concrete. Steel Encased in Concrete or Concrete Filled Tube (CFT)

Cantilevered Steel Framing at Corners

Wuxi Times Square
Structural System Plan

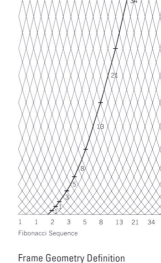

34

21

13

8

5

3

1   1   2   3   5   8   13   21   34   =   88
Fibonacci Sequence

Frame Geometry Definition
Based on Fibonacci Sequence,
Wuxi Times Square

Elevation Drawings of
Evolution of Structural System,
Wuxi Times Square

Wuxi Times Square,
Wuxi, China

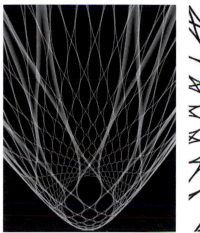

Exterior Frame, Wuxi Times Square,
Wuxi, China

and with the initial numbers in the sequence defined with binary digits of 0
and 1,

$$F_0 = 0$$
$$F_1 = 1.$$

So, for the first number in the sequence ($n = 1$):
$$F_1 = F_{(1-1)} = F_0 = 0$$
For the second number ($n = 2$):
$$F_2 = F_{(2-1)} + F_{(2-2)} = F_1 + F_0 = 0 + 1 = 1$$
For the third number ($n = 3$):
$$F_3 = F_{(3-1)} + F_{(3-2)} = F_2 + F_1 = 1 + 0 = 1$$
For the fourth number ($n = 4$):
$$F_4 = F_{(4-1)} + F_{(4-2)} = F_3 + F_2 = 1 + 1 = 2$$
For the fifth number ($n = 5$):
$$F_5 = F_{(5-1)} + F_{(5-2)} = F_4 + F_3 = 2 + 1 = 3$$
For the sixth number ($n = 6$):
$$F_6 = F_{(6-1)} + F_{(6-2)} = F_5 + F_4 = 3 + 2 = 5$$
For the seventh number ($n = 7$):
$$F_7 = F_{(7-1)} + F_{(7-2)} = F_6 + F_5 = 5 + 3 = 8$$
And so on …

Despite a seemingly infinite variety and diversity of plant growth,
nature employs only three fundamental ways of arranging leaves on a stem.
The first is disticious like corn, the second is decussate like mint, and the
third is spiral phyllotaxis, which represents 80% of the higher order plants

where the rotation angle between leaves is the golden angle of 137.5° (360°/$\emptyset^2$). With this spiral growth, no future leaf overshadows a predecessor, allowing each leaf to receive maximum sunlight and rain—all based on the binary numbers of 1 and 0.

The gravity load-resisting system of the Wuxi Times Square Tower consists of conventional reinforced concrete, structural steel, or a combination of the two. The exterior frame, designed to resist lateral loads only, is connected through floor diaphragms to the interior gravity frame. The geometry of the perimeter frame is defined by the Fibonacci Sequence, with bracing members more vertical with greater spacing at the top of the structure, where cumulative lateral forces requiring resistance are smaller, and more horizontal with tighter spacing near the base, where cumulative lateral forces are largest and so require greatest resistance.

10.5.3   Genetic Algorithms

The Al Sharq Tower is to be located in Dubai, United Arab Emirates. The plan form of the structure is based on nine adjoining cylinders. Since traditional perimeter columns are not desired, a cable-supported perimeter is implemented. Early efforts by the architectural and structural design teams to generate an aesthetically appealing profile followed classic geometric definitions such as that of a helix.

The 102 story residential tower has a 39 m × 39 m (128 ft × 128 ft) floor plan, a height of 365 m (1200 ft), and therefore an aspect ratio nearing

Al Sharq Tower, Dubai, UAE

Darwinian Evolution

DNA Strand

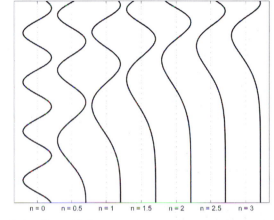

n = 0   n = 0.5   n = 1   n = 1.5   n = 2   n = 2.5   n = 3

Various Tapered Helices

10:1. The proposed structural system consists of reinforced concrete systems with perimeter spiraling high-strength galvanized steel cables. The lateral system is composed of intersecting sets of parallel shear walls and perimeter high-strength galvanized steel cables. The perimeter cable system consists of approximately 70 kilometers (44 miles) of high-strength galvanized steel cables. Initial cable profiles suggested a helical formulation for each. The initial helix definition is propagated to each perimeter cylinder.

10.5.3.1 Cable Profile Influenced by Genetics

Observation of principle stress as a means of form generation is well established in nature. This is demonstrated by the growth patterns of nautilus shells, the fiber-reinforcement of palm tree branches, and the structure of bones which mimic the flow of applied forces. Principle stress observation is undertaken for the preliminary identification of the optimal cable profile for the Al Sharq Tower.

An investigation of principle stresses over the building's perimeter skin is conducted to determine how the form of the building might react to lateral loads. Analysis results show that for the bundled cylinder plan the corner modules exhibit vertical tension (or compression) at the base and transition to 45° (shear) near the top.

Since cables are tension-only members, their most efficient orientation is in alignment with the direction of principle stress. Thus, a cable profile based on observed principle stress trajectories is needed. To define the transition of principle stress trajectories, a modified helical definition is employed. The modified helical definition is shown in Equations 1 through 3, where z_{Total} is the total height of the building, and n is an adjustable parameter that defines the rate of pitch transition over the height of the structure. The ratio of current to total height raised to the power n alters the cable pitch as a function of tower height. A value of $n = 0$ yields no transition of pitch, $n = 1$ yields a linear transition from vertical to 45° over the height of the structure, $n = 2$ yields a parabolic transition, etc. From observation of the principle stresses, it is determined that the transition of principle stress orientation over the face of the windward corner modules is:

$$X(z) = r \cos(t) \tag{1}$$

$$Y(z) = r \sin(t) \tag{2}$$

$$t = z \left(\frac{z}{z_{Total}} \right)^n \tag{3}$$

approximately parabolic ($n = 2$). Thus, a parabolic-tapered helix definition could be used to fully define the cable profile at each perimeter cylinder over the height of the structure. As can be observed, the illustrated parabolic-taper helix definition closely matches that of the corner module of the windward face of the principle stress contours.

Observation of principle stresses at the building perimeter for the determination of cable profile is reasonable if the exterior were monolithic and homogenous. The perimeter is actually a series of discrete tension-only cables. Thus, the principle stress investigation may provide a rational basis for global-perimeter load paths, but further investigation is needed to determine an optimal cable profile.

10.5.3.2   Cable Profile Optimization Using Genetic Algorithms

With a rational basis for perimeter load paths established, further investigation is sought to develop an efficient cable profile for the resistance of lateral loads. Improved performance is pursued through the employment of a genetic algorithm (GA) optimization routine. GA optimization considers a pitch which varies over the height of the tower. In what follows, a general description of the employed GA is provided.

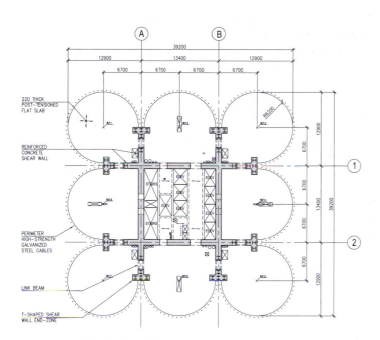

220 THICK
POST-TENSIONED
FLAT SLAB

REINFORCED
CONCRETE
SHEAR WALL

PERIMETER
HIGH-STRENGTH
GALVANIZED
STEEL CABLES

LINK BEAM

T-SHAPED SHEAR
WALL END-ZONE

STAIRS
ELEV
ELEV
ELEV
ELEV
ELEV
STAIRS
ELEV

39200
12900 13400 12900
6700 6700 6700 6700
12900 6700 6700 6700 12900
13400 39200

Floor Framing Plan, Al Sharq Tower,
Dubai, UAE

Elevation, Al Sharq Tower,
Dubai, UAE

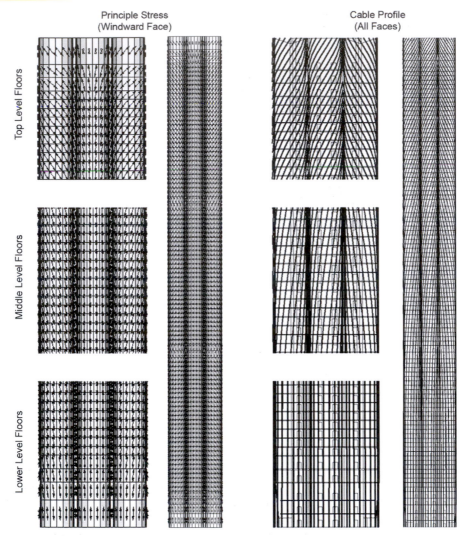

Principle Stress
(Windward Face)

Cable Profile
(All Faces)

Top Level Floors

Middle Level Floors

Lower Level Floors

Principle Stress Analysis and Cable Profile,
Al Sharq Tower, Dubai, UAE

Genetic algorithms have been used in a wide range of applications for improved performance in numerous trades such as the aerospace, automobile, and medical industries. This simple, yet robust, algorithm facilitates multi-variable and multi-objective searches in large, often poorly defined, search spaces. Early investigations of evolutionary algorithms were conducted by Holland (1975) and inspired by observations made by Darwin (1859). GA is a heuristic optimization method which utilizes trial-and-error of mass populations as a basis of optimization. To demonstrate GA concepts, a simple truss optimization problem is illustrated in the following text.

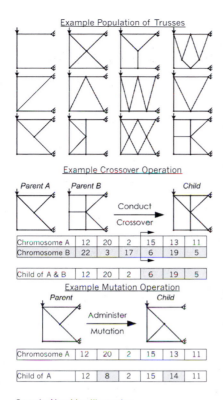

Example Population of Trusses

Example Crossover Operation

Parent A   Parent B   Child

Conduct
Crossover

| Chromosome A | 12 | 20 | 2 | 15 | 13 | 11 |
| Chromosome B | 22 | 3 | 17 | 6 | 19 | 5 |

| Child of A & B | 12 | 20 | 2 | 6 | 19 | 5 |

Example Mutation Operation

Parent   Child

Administer
Mutation

| Chromosome A | 12 | 20 | 2 | 15 | 13 | 11 |

| Child of A | 12 | 8 | 2 | 15 | 14 | 11 |

Genetic Algorithm Illustration

For GA optimization to begin, an initial population must first be generated. A population is a group of candidate-solutions. For the example truss problem, a population would consist of a set of potential truss configurations. Each truss would have a different member configuration but the loading and boundary conditions would be the same.

With an initial population generated, candidate-solutions are evaluated. Their fitness, or score, is determined by a fitness function. For this example, a truss's fitness is the sum of normalized deflection and normalized weight. This GA is a minimization algorithm; thus the sum of the normalized values is taken for the fitness. Both deflection and weight must each be normalized to minimize bias in the fitness score. Analysis software can be used to quickly determine the deflection and weight of each truss in the population. Increased weight and deflection increase the fitness score of a candidate-truss and therefore diminish its chances of being selected by the GA for inclusion in future generations.

The initial population is the first parent population and is used to generate the child population. The child population is a new set of candidate-solutions which are derived from the parent population. The child population

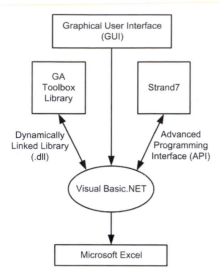

Implementation of Genetic Algorithm

is to be the same size as the parent population. Each member of the child population is to be generated using GA operators. Parameters to be optimized by the GA are contained in a vector of values termed a chromosome.

The first type of GA operation is called "crossover." A crossover operation takes two parents and combines characteristics from each parent to form a child. The second type of GA operation is called "mutation." A mutation operation takes one parent and alters one or more characteristics of the parent to form a child.

After the child population is generated, each child is evaluated and fitness determined. Next, parent and child populations are combined into a single pool of candidate-solutions. The pooled set is ranked according to each member's fitness score. For the truss example problem, the truss with the lowest fitness score is considered best and the truss with the highest fitness score is considered worst. With the pooled parent and child populations ranked, the top 50% are elected to be the parent population for the next generation. The remaining trusses are discarded.

To implement GA for the optimization of the Al Sharq cable filigree, several tools are needed. Visual Basic .NET is a general purpose programming environment well-suited for conducting GA operations, interaction with finite element software, and the collection of results. Finite element analysis software Strand7 is utilized for the analysis of GA-generated cable profiles.

With the concepts of the genetic algorithm described, its application to the optimization of the cable filigree of the Al Sharq Tower is now considered. GA operations are to optimize the cable pitch at each floor. As already observed in the principle stress analysis, optimal cable pitch may

$$J_1 = \frac{1}{(CableArea * TotalCableLength * RoofDrift)}$$

Early Top Performer   **Mid-Run Top Performer**   **Final Top Performer**

Diameter = 29mm   Diameter = 29mm   Diameter = 15mm
Spacing = 8 cables per   Spacing = 6 cables per   Spacing = 6 cables per
         half circle   half circle   half circle
Pitch = 45 deg.   Pitch = 35 deg.   Pitch = 35 deg.
Roof Drift = 771mm   Roof Drift = 831mm   Roof Drift = 2,667mm

Fitness Score = 7.36   **Fitness Score = 8.93**   **Fitness Score = 10.2**

Summary of Fitness Score Results,
Al Sharq Tower, Dubai, UAE

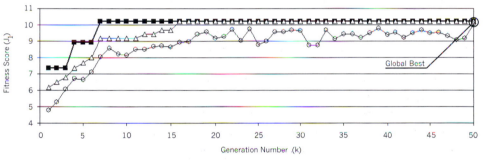

Results of Genetic Algorithm Optimization,
Al Sharq Tower, Dubai, UAE

vary over the height of the tower. With this in mind, GA optimization allows the pitch at each floor to be varied. Thus, 102 variables are concurrently optimized. Fitness function is the normalized roof drift.

A total of 500 generation-cycles are conducted with a population size of 10, thus evaluating a total of 5000 potential cable filigree configurations. Fitness scores steadily improve until approximately generation 350. The top performing solution from generation 500 reveals a cable profile which is very similar to the parabolic profile determined in the previously discussed principle stress-based cable profile study.

Final Interpreted Cable Profile,
Al Sharq Tower, Dubai, UAE

Early efforts to determine an efficient and rational cable profile in response to lateral loads have yielded a cable profile derived from the observation of principle stress contours and confirmed by GA optimization. The optimal cable profile follows a parabolic-helical definition ($n = 2$) which closely matches the principle stress contours observed—vertical at the base transitioning to 45° at the top.

GLOSSARY

CHAPTER 1 PERSPECTIVE

1. **Unreinforced masonry walls**—walls consisting of brick, concrete block or similar materials that do not contain reinforcing steel. May contain cement grout and/or mortar.
2. **Curtain wall system**—an exterior wall system used to enclose the structure and protect against natural elements such as wind, moisture, solar, etc. Wall system does not provide structural support for the base building frame.
3. **Clad structural skeletons**—systems such as frames or walls that provide primary support for the building structure that are clad or covered with exterior curtain wall systems.
4. **Hand calculation techniques**—manual calculations without the use of computers considering the fundamental principles of physics and mathematics.
5. **Force distribution in indeterminate structures**—the distribution of forces due to gravity or lateral loads (i.e. wind or seismic) within a structural frame. Indeterminate structures are those that cannot be defined by the fundamentals of engineering statics. Statics refers to a state of equilibrium where no bodies are in motion and forces are offset or counterbalanced. The stiffness of members must be considered in defining the distribution of forces within the structure.
6. **Well-defined and understandable load paths**—defining structures where a clear load path is defined by one member supporting another due to gravity or lateral loads. A lateral load-resisting system that resists lateral loads through a clearly understood load path from superstructure to foundations.
7. **Higher compressive strengths**—compressive strength of concrete typically achieved with increased amounts of cement in concrete mix. Determined by destructive testing of unreinforced cylinders or cubes. Typically expressed in load per area of material and typically specified/tested at 28 or 56 days after casting.

CHAPTER 2 SITE

1. **Code-defined**—requirements specifically defined by the governing code for design.
2. **Reinforced concrete**—the mixture of cement, sand, aggregate, water, admixtures including accelerators, plasticizers, retarders etc., and reinforcing steel or post-tensioning tendons. Accelerators speed curing (the chemical process of hydration between the cement and water) of the placed mix for conditions such as cold-weather placement of concrete. Retarders slow curing of the placed mix for conditions such as hot-weather placement of concrete.
3. **Ductility**—the ability of the structural system to dissipate energy without failure (i.e. structural steel permanently deforming through bending without fracture).
4. **Detailing**—special structural design and construction requirements used to manage applied loads, allowing primary structural members to fully utilize their strength and stiffness properties. Typically associated with connections within a building frame and are critical in the design of seismically resistant structures.
5. **Redundancy**—providing multiple load paths for forces to travel within the structure. If a member or connection loses its strength due to overload, other members and/or connections will participate in resisting the applied forces and reduce potential for localized or overall progressive structural collapse.

6. **Inertial forces**—forces generated within the structure resulting from applied motion from the ground to the foundations during a seismic event.
7. **Outrigger truss system**—structural steel trusses used to interconnect the core of the structure to the outside frame or columns to resist applied lateral loads.
8. **Direct positive pressure (windward faces)**—pressure created from the velocity of wind on a surface or surfaces perpendicular to the direction of the wind.
9. **Natural period of vibration**—the time required for a structure to complete one full cycle of vibration, referring to the first, primary or dominant mode of vibration when given a horizontal (or vertical) displacement.
10. **Wind velocity**—the speed at which wind travels relative to an at-rest condition.
11. **Pressure tap modeling**—devices or taps included in the physical model for the tower designed to read actual pressure applied to local building surfaces during wind tunnel testing.
12. **Force-balance structural modeling**—block model generally replicating building shapes placed in expected wind environment and boundary layer that includes neighboring terrain/building structures. Bending moments due to applied wind loads are measured and applied to an analytical model of the structure that includes structural member stiffness, mass, and expected damping.
13. **Aero-elastic structural modeling**—model used in wind tunnel testing that incorporates actual building structure properties including shape, mass, stiffness, and damping. Used to directly measure displacements and accelerations during the wind tunnel testing.
14. **Rational wind tunnel studies**—wind studies using modeling techniques that consider actual building properties, site conditions, and historical wind data.
15. **Serviceability including drifts and accelerations**—considerations for the performance of structures in addition to strength primarily related to imposed lateral loads from wind. Drift is the displacement of the structure (between floors or over the height of the structure) due to imposed lateral loads. Exterior wall systems, building partitions, elevator hoistways etc. must be designed for this displacement. Accelerations result from imposed lateral loads on the structure and must be considered for the perception of motion by building occupants.
16. **International Building Code 2006 (IBC 2006)**—a model building code developed by the International Code Council that has been adopted by most jurisdictions in the United States.
17. **Static**—having no motion, objects being acted on by forces that are balanced.
18. **Dynamic**—having motion, applied forces result in motion. In the case of structures exposed to wind forces, it is the response of the structure caused by changes in fluid (air) flow.
19. **Vortex shedding**—the unsteady flow that takes place when a fluid (air) passes an object and creates pockets of alternating low pressure vortices behind the object or downstream of the object. The object tends to move toward areas of low pressure.
20. **Base ten logarithm**—the logarithm of a given number to a given base (10) is the power or exponent to which the base must be raised to produce the number.
21. **Maximum amplitude of oscillation**—the maximum magnitude of change in the oscillating variable, ground motion.
22. **Seismometer**—instrument used to measure and record motions of the ground.
23. **Radiated energy**—during an earthquake stored energy is transformed and results in cracks/deformations in rock, heat, and radiated energy. Represents only a small

fraction of the total amount of energy transformed during an earthquake and is the seismic energy registered on seismographs.

24. **Peak ground (or maximum) acceleration**—the acceleration of a particle on the ground as a result of ground motion.

25. **Oscillatory response**—repetitive variation of an object in motion.

26. **Gravitational acceleration**—the acceleration due to gravity (9.81 m/s^2, 32.2 ft/s^2).

27. **Resonance**—the tendency of the structure to oscillate at larger amplitudes at some periods or frequencies than at others. When a vibrating structure is displaced at a period close to its natural period, accelerations may increase significantly by as much as four or five times.

28. **Zero damping**—a system that is absent of any effect that tends to reduce the amplitude of oscillations in an oscillatory system.

29. **Response spectrum (spectra)**—since seismic ground motion produces transient rather than steady state input, the response spectrum is a plot of the peak response (acceleration, velocity, or displacement) of a series of oscillators with varying natural frequencies or periods that are forced into motion by the same base vibration or shock. The resulting plot is used to determine the response of a structure with a given natural period of vibration, assuming that the system is linear (no permanent deformation of structure under load). Response spectra can be used to assess multiple modes of oscillation (multi-degree of freedom systems that typically include buildings because of multiple locations of mass—i.e. floors), which are only accurate for low levels of damping (most buildings). Modal analysis is performed to identify the modes of vibration and the response for each of those modes can be obtained from the response spectra. Each peak response is then combined for a total response. The typical combination method of these responses is through square root of the sum of the squares (SRSS) provided that the modal frequencies are not close.

30. **Modal analysis**—the study of the dynamic properties of structures when subjected to vibration-induced excitation.

31. **Spectral acceleration**—approximately the acceleration experienced by a building, as modeled by a particle mass on a massless vertical rod having the same natural period as the building. When the mass-rod system is moved or "pushed" at its base using a seismic record, assuming a certain damping to the mass-rod system (typically 5%) one obtains a record of particle motion.

32. **Spectral velocity**—the rate of change (derivative) of the displacement record with respect to time.

33. **Spectral displacement**—is approximately the displacement experienced by the building as modeled by a particle mass on a massless vertical rod having the same natural period as the building. Based on the seismic record, the maximum displacement can be recorded.

34. **Initial loading**—loads typically associated with the dead weight of the superstructure.

35. **Pre-consolidated**—soils that have deformed over time as a result of load usually associated with clay, where the soil "squeezes" over time when subjected to compressive loads.

36. **Creep effects**—the tendency of a solid material (soil in this case) to slowly move or deform permanently under the influence of stresses due to sustained loads.

37. **Spread or continuous footings**—foundation system that distributes load from the superstructure to the soil system. Usually comprised of reinforced concrete with spread footings square or rectangular in shape and continuous footings having a specific width but continuing along under multiple columns or walls. Continuous footings are typically used under perimeter foundation walls.

38. **Mat foundations**—foundation system that distributes load from the superstructure to the soil system typically over an area much larger than spread footings. Usually comprised of reinforced concrete and supports multiple vertical load carrying elements such as columns or walls.

39. **Piles or caissons**—vertical or sometimes sloped load-carrying elements used to impose load on suitable soil layers below the surface of the ground. Loads are typically transferred by end bearing and skin friction between the pile or caisson and the soil. The lengths of piles or caissons vary based on applied loads and soil conditions. Piles or caisson lengths can vary from a few feet to several hundred feet.

40. **Soil stratum or layers**—layers of soil having specific thicknesses and geotechnical/structural characteristics.

41. **Pile cap**—structural element used to transfer load from the superstructure to pile systems. Usually comprised of reinforced concrete.

42. **Skin friction**—the friction developed between the surface of a pile or caisson and the neighboring soil.

43. **H-piles, precast piles, steel pipe piles**—driven piles consisting of structural steel H-shaped sections, precast concrete, and open or closed end steel pipes.

44. **Straight-shafted or belled caissons**—a caisson may have a constant cross-section along its full length or may have it along a majority of its length with an enlarged section at its base shaped like a bell.

45. **Stiff clay (hardpan)**—clay having high compressive resistance with little susceptibility to consolidation or creep over time. The city of Chicago is known for having this soil condition and has many tall buildings supported by caissons that bear on this soil layer.

46. **Slurry walls**—foundation system typically installed with bentonite slurry. Trenches are typically dug from the surface with bentonite installed to keep neighboring soil from collapsing into the trenches. Bentonite slurry has higher density than neighboring soil to prevent soil from collapsing and is displaced from the trench during the concrete placement process. Walls are typically installed in panels or segments with shear keyways placed in between.

47. **Shear keyways**—continuous vertical notches placed in slurry walls to transfer shear loads imposed from soil from one panel or segment to the next.

48. **Compressible (soil) layers**—soil layer susceptible to displacement when subjected to compression loads.

49. **Settlement and differential settlement**—displacement of soil systems when subjected to load. Settlements may differ across the foundation system due to unsymmetrical loads or varying soil conditions.

50. **Pressure grouting**—a technique used to strengthen soil layers. Cement grout is typically injected into the soil to increase strength and limit settlement.

51. **Subgrade moduli**—the relationship of stress and strain in soil. Used to determine soil stiffness and susceptibility to settlement.

52. **Out-of-plumb**—alignment that varies from a pure vertical condition.

CHAPTER 3 FORCES

1. **kPa**—kilopascal.
2. **psf**—pounds per square foot.
3. **Gravity loads**—loads due to the self-weight (based on density of the material used) of the structure or superimposed loads (dead or live) on the structure.

4. **Dead load**—load that is constant over time, typically referring to the structure's self-weight.

5. **Superimposed load**—permanent load that is constant over time. Loads are due to the weight of exterior wall systems, partitions, and the like.

6. **Live load**—temporary loads typically of short duration. Typically includes occupancy loads or snow loads.

7. **Lateral loads**—laterally applied loads typically due to temporary events such as wind or seismic activity.

8. **Vertical force distribution**—distribution of lateral forces typically due to wind or seismic loadings over the height of the structure. Forces can be applied as distributed loads or point loads at floor diaphragms.

9. **Earthquake force**—force experienced by structure due to ground motions from seismic events.

10. **Design spectral response acceleration**—equal to 2/3 of maximum considered spectral response acceleration for a site resulting from a 475-year seismic event.

11. **Maximum considered spectral response acceleration**—the maximum considered spectral acceleration for a site resulting from a 2475-year seismic event.

12. **Reliability/redundancy factor**—a factor used in the calculation of earthquake forces that considers the reliability/redundancy of a specific structural system.

13. **Overstrength factor or seismic force amplification factor**—a factor applied to structural elements that are expected to remain elastic during the design ground motion to ensure system integrity.

14. **Seismic base shear**—the force applied to the base of a structure due to seismicity. This force is distributed over the height of the structure.

15. **Seismic response coefficient**—a coefficient based on spectral response acceleration, a structural system-dependent response modification factor, and an occupancy importance factor. This coefficient is applied to the weight of the structure to determine the seismic base shear.

16. **Effective seismic weight**—dead and live load specifically considered in seismic analysis.

17. **Occupancy importance factor**—a factor used to recognize structures with occupancies that require special considerations (for example acute care hospitals).

18. **Response modification factor**—a modification factor that accounts for a specific structural system. The factor increases with system ductility. For instance, a steel moment-resisting frame has a higher response modification factor than a steel braced frame.

19. **Fundamental period of vibration**—the longest period (or lowest frequency) of vibration for the structure. Also referred to as the first or primary mode of vibration.

20. **Seismic zone factor**—a factor associated with mapped zones of anticipated ground accelerations.

21. **Horizontal force distribution**—distribution of horizontal story forces along the height of the structure.

22. **Bending moment distribution**—applied bending moments along the height of the structure.

23. **Inter-story drifts**—relative displacement of one level of the structure relative to another due to applied load.

24. **Inelastic drift**—displacement of the structure when applied loads cause permanent deformations.

25. **Elastic drift**—displacement of the structure when applied loads do not cause permanent deformations and allow the structure to return to its original undeformed position.
26. **Column tributary area**—floor-framing area supported by vertical columns.
27. **Axial load**—load applied along the axis of a member or system.
28. **Shear load**—load applied across the axis of a member or system.
29. **Bending moments**—load applied to a member or system the causes the structure to bend.

CHAPTER 4 LANGUAGE

1. **Force flow**—the internal flow of force throughout a structure.
2. **Wide-flanged beam analogy**—force distribution analogy comparing a wide-flanged beam to columns located in the plan of a tall building.
3. **Prestress**—placement of loads on a structure to create an initial state of stress; usually introduced to offset other applied loads.
4. **Concentric braced frames**—diagonal frames in a structure that share a common joint without any offset.
5. **Diagonal braces**—braces within the structure placed diagonally to resist loads.
6. **Bundled steel tubular frame**—an assembly and interconnection of individual tubular structural steel frames.
7. **Screen frames**—frames designed to resist structural load while providing other benefits such as screening or shade from the sun to control building temperatures.
8. **Shear wall buttressed core**—a structural system that combines a central shear wall core with other buttressing shear wall elements.
9. **Composite**—in tall building construction usually refers to the combination of steel and reinforced concrete.
10. **Mega-column**—column of large scale typically used in structural systems where gravity loads are concentrated, or in an outrigger truss system where high axial load resistance is required, or a combination of both.
11. **Steel-plated core**—structural steel used in core systems as an alternative to a reinforced concrete shear wall system.

CHAPTER 5 ATTRIBUTES

1. **Strength**—capacity of a material or structure to resist applied forces.
2. **Limit state**—the application of statistics to determine the level of safety required for the design of structural members or systems.
3. **Redundancy**—the duplication of a critical structural component to increase the reliability of the overall structural system.
4. **Load and resistance factor design (LRFD)**—synonymous with limit state, the application of statistics to determine the level of safety required for the design of structural members or systems. Most commonly known for the structural steel design code that superseded allowable stress design.
5. **Allowable stress design**—the capacity of a structural material is based on the allowable load that the material could resist without permanent or plastic deformation. Stresses due to service loads do not exceed the elastic limit of the material. Also known as permissible stress design.

6. **Probability of exceedance**—the statistical probability of exceedance relative to a specified return period.
7. **Serviceability**—performance characteristics of the structure including but not limited to drift, damping, acceleration, creep, shrinkage, and elastic shortening.
8. **Wind-induced motion**—motion induced into a structure by wind forces.
9. **Damping ratio**—an effect that acts to reduce the amplitude of oscillation. Usually referred to as the percent of critical damping. Building elements that provide damping include non-structural components such as partitions, ceilings etc., the structure itself, aerodynamics, and so on.
10. **Strain energy**—the external work done on an elastic member in causing it to distort from its unstressed state is transformed into strain energy.
11. **Acceleration**—the rate of change of velocity over time. In tall building design, wind-induced acceleration is usually expressed in milli-gs or thousandths of the acceleration of gravity.
12. **Creep**—the tendency of material to deform permanently under the influence of stress.
13. **Shrinkage**—a phenomenon that occurs in concrete where volume is reduced through the hydration/drying process.
14. **Elastic shortening**—the reduction of length in a structural element due to load. The element returns to its undeformed shape after the load is removed.
15. **MEP systems**—mechanical/electrical/plumbing systems.
16. **Long-term loads**—sustained loads on a structure such as dead load and permanent superimposed load, for example from exterior wall systems.
17. **Modulus of elasticity**—the linear relationship of stress to strain in a material; the tendency of a material to behave elastically when load is applied to it.

CHAPTER 6 CHARACTERISTICS

1. **Dynamic properties**—natural properties of the structure that include mass and stiffness.
2. **Aerodynamics**—air flow around a solid object of various shapes.
3. **Period**—the time it takes for one full oscillation of a structure; the reciprocal of frequency.
4. **Across-wind motion**—motion of the structure in the direction of the applied wind load. Also known as lift.
5. **Along-wind motion**—motion of the structure perpendicular to the applied wind load. Also known as drag.
6. **Vortex shedding**—the unsteady air flow that takes place in special flow velocities according to the size and shape of the structure. Vortices are created at the back of the structure and detach periodically from either side of the structure. The fluid flow past the object creates alternating low-pressure vortices and the structure tends to move toward those areas of low pressure.
7. **Aspect ratio**—the ratio of the height of a structure relative to its smallest plan dimension at the base.

CHAPTER 7 SYSTEMS

1. **Moment of inertia**—a property of a cross section that can be used to predict the resistance of a structure when subjected to bending or deflection, around an axis that lies in the cross-sectional plane.

2. **Semi-rigid frame**—a moment-resisting steel frame that incorporates connections that have partial fixity allowing for some rotation when loaded.
3. **Rigid frame**—a moment-resisting steel, concrete or composite frame that incorporates fully fixed connections.
4. **Shear truss**—a steel truss typically consisting of diagonal members and located in the core area of the building, designed to resist lateral shear.
5. **Belt truss**—a steel truss used to tie perimeter columns together and typically used to transfer load from outrigger trusses to perimeter columns.
6. **Outrigger truss**—a steel truss used to share load between a steel shear truss or concrete shear wall core and columns to resist overturning caused by lateral loads.
7. **Chevron or k-braced truss**—a steel truss with a geometry that is in the shape of a K with a horizontal orientation; typically used in core or outrigger truss applications.
8. **Concentrically braced frame**—a braced frame where diagonals meet at a common work point.
9. **Eccentrically braced frame**—a braced frame where work points of diagonals are not common at horizontal members.
10. **X-braced frames**—frames that introduce crossing diagonals into the truss system sharing work points; may be used on core, perimeter, belt or outrigger truss systems, among others.
11. **Exterior frame tube or tubular frame or framed tube**—a frame with closely spaced columns having a plan spacing typically similar to the floor-to-floor height.
12. **Bundled frame tube**—a bundling or collection of tubular frames that typically pass through the interior of the structure; belt trusses are typically used at building geometry transitions.
13. **Cellular concept**—placing frames in plan to form the geometry of cells in the interior of the structure.
14. **Exterior diagonal tube**—diagonal structural members incorporated into a tubular system at the perimeter of the structure.
15. **Superframe**—a frame located at the perimeter of the structure that incorporates three-dimensionally placed members typically including diagonals.
16. **Diagonal mesh tube frame**—smaller, more repetitive diagonal structural members incorporated into the structure typically at the perimeter.
17. **Shear wall**—reinforced concrete wall system designed to resist shear caused by lateral loads.
18. **Frame - shear wall**—a combination of a frame and shear wall.
19. **Tube-in-tube**—the combination of frames typically in the center core area and at the perimeter of the structure. The frames typically consist of closely spaced columns having a plan spacing similar to the floor-to-floor height.
20. **Modular tube**—a tubular frame that is included both at the perimeter of the structure and through its interior.
21. **Diagonal concrete braced frame**—a braced frame in reinforced concrete where panels are infilled between frame columns to form diagonal patterns.
22. **Belt shear wall stayed mast**—a central reinforced concrete shear wall is interconnected with perimeter columns or frames with outrigger trusses or walls and interconnected to perimeter belt walls or trusses.
23. **Mega-core shear walls**—large reinforced concrete cores usually located in the central core area and designed to resist a majoring of load with or without interconnection with perimeter columns or frames through an outrigger system.
24. **Axial stiffness**—stiffness of axial load-resisting elements within the structure;

important to the effectiveness of outrigger trusses interconnected between the core and columns.

25. **Diagonal concrete mesh tube frame**—a reinforced concrete frame made up of smaller, more repetitive diagonal members; typically located at the perimeter of the structure and has cellular characteristics.

26. **Composite**—typically associated with the combination of structural steel and concrete in a structural member or system.

27. **Composite frame**—a frame that includes columns and beams comprised of both structural steel and reinforced concrete.

28. **Shear wall - steel gravity columns**—a lateral load-resisting reinforced concrete shear wall core combined with steel columns that only resist gravity loads and provide no lateral load resistance.

29. **Shear wall** - composite frame—a reinforced concrete shear wall core combined with a frame that includes columns and beams comprised of both structural steel and reinforced concrete.

30. **Service area**—typically the area within the core that includes stairs, elevators, mechanical rooms, and the like.

31. **Stiffness**—the product of a material's modulus of elasticity and moment of inertia or axial area.

32. **Net tension**—the resulting force on a member or system where tensile forces are greater than compressive forces due to axial loads or bending.

33. **Uplift**—net upward force on a structural element or system; will cause upward displacement if not restrained.

34. **Overturning**—the tendency for the structure to overturn due to applied loads.

35. **Moment-resisting frame**—a frame with fixed beam-to-column joints designed to resist lateral loads.

36. **Floor-to-floor height**—the overall vertical dimension of the structure from floor slab at one level relative to the floor slab at a neighboring level.

37. **Shear lag**—inefficiencies in transferring forces to columns on the faces of the structure perpendicular to the application of load, typically in a tubular or mesh tube frame.

CHAPTER 8 NATURE

1. **Screen frames**—structural frames comprised of structural steel, concrete, or composite that are used to resist lateral loads and to shade the structure or the like; may have infill panels that are not symmetrical.

2. **Mega-frame concept**—the use of a frame that typically extends over multiple stories of the structure.

3. **Stiffened screen frame**—frame with infill panels that are introduced to provide additional stiffness to resist lateral loads.

4. **Prestressed frames**—the use of prestressing to balance load in frames providing strength and deflection control; typically included in reinforced concrete structures but could be included in steel or composite structures.

5. **Rigid diaphragms**—horizontal-framing systems typically consisting of concrete slabs used to interconnect the structural system together.

6. **Growth patterns**—patterns of natural growth observed in nature having structural characteristics.

7. **Quadratic formulation**—a polynomial equation to the second degree; used in structural engineering mechanics.

8. **Bamboo**—a plant having a mathematically predictable growth pattern, consistent structural properties, and fast natural growth.
9. **Culm**—the stalk or trunk of bamboo, typically hollow except at nodal or diaphragm locations.

CHAPTER 9 MECHANISMS

1. **Mechanism**—a structure having joints that allow moving without permanent deformation of materials.
2. **Beam-to-column**—a joint that describes the connection of a beam to a column.
3. **Reduced beam section (RBS) or dogbone connection**—a connection that incorporates a reduced cross-sectional area in a frame beam just outside of the column face but eliminating part of the flanges.
4. **Relative displacement**—unequal displacement between two structural members, typically between columns or walls in a floor.
5. **Pinned joints**—the use of pins to eliminate moment resistance in a joint.
6. **Pinned-truss concept**—the use of pins to allow for movement in a truss system when subjected to relative end displacements.
7. **Cyclic behavior**—structures subjected to loads or displacements that start in one direction and then reverse to the other.
8. **Mill scale**—the surface of steel left after a typical milling process.
9. **Faying surface**—an interface surface between two or more materials that has a characteristic coefficient of friction.
10. **Shim**—a material used between two other materials that is required to take up dimension or to provide a predictable coefficient of friction.
11. **Belleville washers**—compressible washers used to maintain bolt tension after sliding in a joint has occurred.
12. **Direct tension indicators (DTIs)**—washers used to predict bolt tension through the deformation of tabs incorporated into the washers.
13. **Pin-Fuse Joint®**—a patented seismic frame device that uses friction fuse to dissipate energy during an earthquake; uses a pin, curved steel plates, brass shims, and high strength steel bolts.
14. **Slip-critical**—typically refers to bolted connections where the threshold of slip is important.
15. **Friction-type**—a connection that relies on friction to determine capacity.
16. **Pin-Fuse Frame™**—a patented seismic frame device that uses friction fuses to dissipate energy during an earthquake; uses pins, circular bolt arrangements for beam end connections, brass shims, long-slotted holes in diagonal braces, and high strength steel bolts.
17. **Link-Fuse Joint™**—a patented seismic frame device that uses friction fuses to dissipate energy during an earthquake; uses pins, butterfly link-beam steel plate connection, brass shims, and high strength steel bolts.
18. **Optimal structural typology**—conceiving of structural systems as membranes and performing optimization analyses to determine regions where the greatest structural material density is required based on applied load and support conditions.
19. **Logarithmic spiral**—a spiral form that is mathematically defined using logarithmic theory.
20. **Michell Truss**—the mathematical derivation by Anthony Michell defining the geometry of the perfect cantilever subjected to a lateral load with two points of support.

CHAPTER 10 ENVIRONMENT

1. **Enhanced seismic systems**—structural systems that introduce technologies that lead to better performing structures in seismic events where higher performance and less damage are achieved.

2. **Sustainable Form Inclusion System (SFIS)™**—a patent pending form inclusion system that introduces post-consumer waste products into structures to reduce waste while also minimizing structural materials through a reduction in mass.

3. **Environmental Analysis Tool™**—a patent pending calculator used to determine the carbon footprint, cost-benefit, and life cycle of a structure.

4. **Fragility curves**—mathematical curves defining the performance of a structural system when subjected to large displacements and loads; used for risk assessments of structures.

5. **Life cycle**—the estimated life of a structure given site and loading conditions.

6. **Cost-benefit analysis of structures**—the evaluation of first cost relative to long-term cost of a structural system considering devices designed to provide better structural performance.

7. **Probable Maximum Loss (PML)**—the estimated amount of financial loss expected after a prescribed earthquake expressed as a whole number from 0 to 100 (representing the loss percentage of total building value).

8. **Emergence theory**—emergence or self-organization is the interaction between simple common elements having singular and common characteristics, each functioning according to its own simple rules, resulting in complex behavior, without an obvious central controlling force.

9. **Optimization**—analysis techniques used to calculate the minimum amount of material required for particular loadings and boundary conditions.

10. **Helical formation**—a geometric definition of a form based on the helix.

11. **Genetic algorithm**—search technique used in computing to find exact or approximate solutions to optimization and search problems based on the concepts of genetics.

12. **Fibonacci sequence**—rooted in logic formations created by binary numbers; it is a series of numbers starting with 0 and 1 where the subsequent number in the sequence is equal to the sum of the previous two numbers.

13. **Binary digits**—the digits 0 and 1.

14. **Boundary conditions**—support conditions of the structure that must be concurrently considered with the application of load.

REFERENCES

Ambrose, J. and Vergun, D. *Simplified Building Design for Wind and Earthquake Forces*, 2nd edition, John Wiley & Sons, 1990.

American Society of Civil Engineers (ASCE 88), formally American National Standards Institute (ANSI 58.1), *Minimum Design Loads for Buildings and Other Structures*, 1988.

American Society of Civil Engineers (ASCE 7-10), *Minimum Design Loads for Buildings and Other Structures*, 2010.

Chinese Building Code, *JGJ 99-98 Technical Specification for Steel Structures of Tall Buildings*, 1998.

Chinese Building Code, *GB 50011-2001 Code for Seismic Design of Buildings*, 2001.

Chinese Building Code, *JGJ 3-2002 Technical Specification for Concrete Structures of Tall Buildings*, 2002.

Chinese Building Code, *CECS 230: 2008 Specification for Design of Steel-Concrete Mixed Structure of Tall Buildings*, 2008.

Darwin, C. *The Origin of Species by Means of Natural Selection, or the Preservation of Favoured Races in the Struggle for Life*, John Murray, London, 1859.

Fanella, D. and Munshi, J. *Design of Concrete Buildings for Earthquake and Wind Forces*, Portland Cement Association (PCA), according to the 1997 Uniform Building Code, 1998.

Holland, J. *Adaptation in Natural and Artificial Systems*, University of Michigan Press, Ann Arbor, MI, 48016, 1975.

International Code Council, *International Building Code (IBC) 2006: Code and Commentary, Volume 1*, 2006.

International Conference of Building Officials (ICBO), *Uniform Building Code (UBC)*, Structural Engineering Design Provisions, Whittier, CA, 1997.

Intergovernmental Panel on Climate Change (IPCC), "Climate Change 2007: Synthesis Report," Contribution of Working Groups I, II and III to the Fourth Assessment Report of the Intergovernmental Panel on Climate Change, Core Writing Team, eds. R.K. Pachauri and A. Reisinger, IPCC, Geneva, Switzerland, 2007, p. 104.

Janssen, J.J.A. *Mechanical Properties of Bamboo*, Springer, New York, 1991.

Lindeburg, M. and Baradar, M. *Seismic Design of Building Structures 8th Edition*, Professional Publications, Inc., Belmont, CA, 2001.

National Building Code of Canada (NBC), *Structural Commentary Part 4 – Wind Engineering*, 2005.

Sarkisian, M. *Structural Seismic Devices*, United States Patent Nos. US 6,681,538, 7,000,304, 7,712,266, 7,647,734.

US Green Building Council (USGBC), "Buildings and Climate Change," USGBC Report, 2004.

Willis, C. (ed.), *Building the Empire State*, W.W. Norton & Company Inc., New York, copyright the Skyscraper Museum, 1998.

ACKNOWLEDGMENTS

The Work
The Architects and Engineers of Skidmore, Owings & Merrill LLP

Contributors to the book development
Structural Engineering
Eric Long
Neville Mathias
Jeffrey Keileh
Danny Bentley

Original Graphics
Lonny Israel
Brad Thomas

Publication Coordination/Images
Pam Raymond
Harriet Tzou
Rae Quigley
Justina Szal
Brian Pobuda

SOM Integrated Design Studio—Stanford University
Brian Lee
Leo Chow
Mark Sarkisian
Brian Mulder
Eric Long

Editing
Cathy Sarkisian

Inspiration
Stan Korista

Special acknowledgment to Eric Long for his tireless efforts in helping make this book possible.

THE AUTHOR

MARK SARKISIAN, PE, SE, LEED® AP, is a structural engineer and leads the Structures Studio in Skidmore, Owings & Merrill's San Francisco office. Since joining SOM in 1985, Sarkisian's career has focused on creating inventive engineering solutions carefully integrated into the architecture of some of the world's largest and most complex building projects. He works collaboratively to develop new approaches to building design contemplating natural growth and environmentally responsible structures. He is a visiting lecturer at Stanford University, teaching an integrated studio design class where structural engineering and architecture are considered in parallel, with equal emphasis. He also teaches an integrated studio design class to students from the University of California, Berkeley; California Polytechnic State University, San Luis Obispo; and the California College of the Arts in San Francisco. Sarkisian has written nearly one hundred technical papers on innovative approaches to building design.

INDEX